GENETICALLY ENGINEERED MARINE ORGANISMS

Environmental and Economic Risks and Benefits

GENETICALLY ENGINEERED MARINE ORGANISMS

Environmental and Economic Risks and Benefits

edited by

Raymond A. Zilinskas
University of Maryland Biotechnology Institute
College Park, Maryland U.S.A.

and

Peter J. Balint
University of Maryland Biotechnology Institute
College Park, Maryland U.S.A.

KLUWER ACADEMIC PUBLISHERS
Boston / Dordrecht / London

Distributors for North, Central and South America:
Kluwer Academic Publishers
101 Philip Drive
Assinippi Park
Norwell, Massachusetts 02061 USA
Telephone (781) 871-6600
Fax (781) 871-6528
E-Mail <kluwer@wkap.com>

Distributors for all other countries:
Kluwer Academic Publishers Group
Distribution Centre
Post Office Box 322
3300 AH Dordrecht, THE NETHERLANDS
Telephone 31 78 6392 392
Fax 31 78 6546 474
E-Mail <services@wkap.nl>

 Electronic Services <http://www.wkap.nl>

Library of Congress Cataloging-in-Publication Data

A C.I.P. Catalogue record for this book is available
from the Library of Congress.

Copyright © 1998 by Kluwer Academic Publishers

All rights reserved. No part of this publication may be reproduced, stored in a retrieval system or transmitted in any form or by any means, mechanical, photo-copying, recording, or otherwise, without the prior written permission of the publisher, Kluwer Academic Publishers, 101 Philip Drive, Assinippi Park, Norwell, Massachusetts 02061

Printed on acid-free paper.

Printed in the United States of America

Contents

Foreword   vii

Preface   xi

List of Contributors   xv

Acknowledgments   xvii

Risk Assessment for Uncontained Applications of Genetically Engineered Organisms   1
MORRIS LEVIN

Characteristics of Marine Ecosystems Relevant to Uncontained Applications of Genetically Engineered Organisms   31
JOHN J. GUTRICH, HOWARD H. WHITEMAN, AND RAYMOND A. ZILINSKAS

Analysis of the Ecological Risks Associated with Genetically Engineered Marine Macroorganisms   61
JOHN J. GUTRICH AND HOWARD H. WHITEMAN

Analysis of the Ecological Risks Associated with Genetically Engineered Marine Microorganisms   95
RAYMOND A. ZILINSKAS

Federal and State Regulations Relevant to Uncontained Applications of Genetically Engineered Marine Organisms   139
SUSAN STENQUIST

Economic Analysis of Introduced Genetically Engineered Organisms   181
DIANE HITE AND JOHN J. GUTRICH

Risks and Benefits of Marine Biotechnology: Conclusions and Recommendations   213
PETER J. BALINT, RITA R. COLWELL, JOHN J. GUTRICH, DIANE HITE, MORRIS LEVIN, SUSAN STENQUIST, HOWARD H. WHITEMAN, AND RAYMOND A. ZILINSKAS

List of Acronyms   221

Index   225

Foreword

Rita R. Colwell

Since the publication by Professor Paul Berg and his colleagues calling for a moratorium on genetic engineering research appeared in *Nature* on July 19, 1974, the landscape of biotechnology has been dramatically altered. Advances in molecular biology, recombinant DNA research, and genetic manipulation have resulted in impressive new technologies for transfer of genetic material intra-, inter-, and supra-generically. As a result, a cascade of requests has flowed into regulatory agencies of the United States government over the past 25 years for permission to release genetically engineered organisms into the environment, both for research and development and for commercial purposes.

In the early 1980s, a cabinet council working group was appointed under the direction of the Office of Science and Technology Policy to consider the scientific, legal, and policy issues associated with environmental applications of biotechnology. This panel included representatives from the U.S. Environmental Protection Agency (EPA), National Institutes of Health (NIH), U.S. Department of Agriculture (USDA), and other agencies. A proposal for a coordinated approach to the regulation of biotechnology applications was published in December 1984. The White House Biotechnology Science Coordinating Committee refined these guidelines and provided a framework for federal regulation of biotechnology that was directed at the product and not the process of recombinant DNA. This proposal had the advantage of offering industry a measure of regulatory certainty. International competitiveness was enhanced as U.S. biotechnology firms were able to move forward with commercialization of useful products. The extraordinary expansion of this sector of the U.S. economy, now including more than 1,300 companies capitalized at greater that $80 billion, attests to the success of the industry under the regulations presently in place.

The issue of biotechnology regulation is far from closed, however. The public remains concerned about some aspects of biotechnology applications, notably those involving introductions into the open environment. Four regulatory agencies—USDA, EPA, the Food and Drug Administration (FDA), and the Occupational Safety and Health Administration (OSHA)—presently share oversight responsibility under existing laws. FDA regulates human and animal drugs, medical devices, and biologics and serves as the lead agency for oversight of food and food additives. OSHA oversees workplace hazards. EPA serves as the lead agency for regulating microbial pesticides released into the environment. In addition, USDA, NIH, and the National Science Foundation oversee research activities.

Scientific advances are now leading to the application of genetic engineering technology in marine species, including fish, shellfish, and microorganisms. This book focuses on the implications for risk assessment of introducing transgenic organisms into the marine environment. Existing risk assessment approaches for terrestrial organisms are considered and their suitability for use as tools to assess applications of marine biotechnology is evaluated. Challenges for risk assessment include the potential for escape of genetically engineered organisms leading to the establishment of viable populations, possible transfer of altered genetic material to native marine organisms, and subsequent ecological effects arising from interactions between introduced organisms and marine hosts or other environmental factors. Available risk assessment methods are examined critically, and a model for ecological risk analysis is offered.

The authors conclude that, while methods for terrestrial ecosystem structural analysis are useful in identifying potential risks, actual observation and testing in marine systems is essential for assuring the success and safety of applications of biotechnology in the marine environment. Controlled testing and monitoring procedures need to be developed to detect, identify, and enumerate specific marine microorganisms, predict the fate and effects of genetically engineered micro- and macroorganisms, and assess the genetic stability and potential for genetic transfer of introduced organisms. Clearly, accelerated research in marine ecology and systematics—fields of study that recently have been neglected as unfashionable or irrelevant—is urgently required. Knowledge in these subjects is vital to underpin decisions relating to release of specific transgenic marine organisms.

This volume takes us a significant step forward in preparing to implement credible risk assessment procedures in preparation for the release of bioengineered organisms into the open marine environment.

Preface

Raymond A. Zilinskas and Peter J. Balint

Following early controversy in both the scientific community and society at large concerning the implications of advances in recombinant DNA technology, we now have over two decades of experience supporting the conclusion that genetically engineered organisms (GEOs) are not inherently dangerous to public health or to the environment. To date, no adverse effects of agricultural, medical, or pharmaceutical uses of GEOs have been observed.

Public fears have not entirely dissipated, however. In November 1996, protesters demonstrating against the production and distribution of genetically engineered crops disrupted the United Nations World Food Conference, held that year in Rome. The following month the European Union, responding to widespread consumer resistance, decided to require labeling of imported American corn and soybeans grown from genetically altered seeds, to distinguish these products from crops grown from conventional seeds. The public in the United States has also responded with anxiety to advances in biotechnology. For example, there were negative reactions both to the introduction of a genetically engineered growth hormone designed to enhance milk production in cows in 1993 and to the cloning of an adult sheep in early 1997.

Concerns remain among experts as well. Regulators and biotechnology firms find themselves working with inconsistent oversight procedures, biological scientists see gaps in the science base necessary for credible risk assessment, and policy makers face the challenge of encouraging economic development in the biotechnology industry while simultaneously ensuring public health and environmental safety.

Now GEOs are being designed for application in the marine environment, particularly in aquaculture and bioremediation of pollution. Concerns associated with pharmaceutical and agricultural biotechnology products appear likely to reemerge with development of transgenic fish, shellfish, and marine microorganisms. Stakeholders include the biotechnology industry, the scientific community, regulatory agencies at various levels of government, and groups representing the public interest.

This book—designed to synthesize present knowledge concerning the implications of marine biotechnology and offer recommendations for policy-making—reports the results of two related studies carried out over a four-year period by the Center for Public Issues in Biotechnology (CPIB), University of Maryland Biotechnology Institute. These studies focused primarily on ecological and economic issues. As a result, the authors do not undertake an examination of human health concerns in this volume. Included in the analysis are risk assessment, environmental law, ecology and evolutionary biology, microbiology, and economics. Both projects were supported by the National Sea Grant Program of the U.S. National Oceanographic and Atmospheric Administration (NOAA).

The first of the two studies, which began in 1993, examined biosafety implications of marine biotechnology (Zilinskas et al., 1995). Investigators found that a greater degree of uncertainty will be associated with uncontained field trials of transgenic organisms in the marine environment than has been the case with terrestrial applications. First, it will be difficult, if not impossible, to ensure the biological isolation of GEOs tested in the open ocean. In addition, potential ecological risks are compounded by factors including broader opportunities for dispersal available to marine organisms and the existence in the oceans of pathways for exchange of genetic material that remain poorly understood and have no analogs in the terrestrial environment. For example, recent investigations have strengthened the hypothesis that viruses inhabiting the ocean surface layer and sea floor sediment may facilitate gene transfer in marine ecosystems (Wommack et al., 1992).

A second conclusion drawn from the initial study was that existing risk assessment protocols need to be reevaluated, and supported by further scientific research, if they are to be useful in estimating potential hazards of uncontained applications of marine GEOs. Because adequate information is not currently available on the genetics of fish, shellfish, and microorganisms in marine ecosystems, effects of marine GEOs cannot be estimated with confidence. It will be difficult to determine in advance the probability of escape of test organisms, the likelihood that fugitive populations will establish themselves in new habitats, the dispersal range available to escapees, or the possible ecological effects of established transgenic populations.

The second CPIB project began in 1994 with the aim of conducting a more comprehensive study of the applicability to marine biotechnology of existing laws, regulations, and risk assessment methods. Results included the development of a protocol for ecological risk analysis that takes into account the unique characteristics of marine ecosystems and recommendations for further research to build the knowledge base necessary for scientific risk assessment. The study also estimated costs to industry and research institutions of compliance with proposed risk assessment and risk management procedures and considered the economic benefits that may accrue to industry, and to society, as a result of advances in marine biotechnology, particularly in aquaculture and bioremediation.

In this book, the contributors present results and conclusions from both studies. Chapter 1 provides an overview of the evolution of regulatory risk assessment as applied to biotechnology products. The author considers procedures used by the Environmental Protection Agency (EPA), Department of Agriculture (USDA), and Food and Drug Administration (FDA) to manage risk associated with terrestrial trials of GEOs; discusses how these methods may apply to marine trials; compares U.S. protocols with international risk assessment schemes designed to regulate marine biotechnology; and makes recommendations to safeguard the environment while encouraging the development of marine biotechnology.

Chapter 2 focuses on physical and ecological characteristics of the ocean environment that distinguish it from land and freshwater systems, thereby laying the groundwork for later chapters that assess implications for risk assessment with reference both to macro- and microorganisms. The authors examine physical features of the oceans, including scale, dimensionality, continuity, and movement, and discuss how these properties influence biotic factors, including life history strategies, trophic level complexity, and ocean biogeography.

Chapter 3 builds on the conclusions of previous chapters to provide guidance to those responsible for developing and implementing risk assessment protocols for marine biotechnology. Concentrating on macroorganisms, the authors contrast problems posed by the unique features and complexities of the marine environment with those of terrestrial and freshwater ecosystems and propose a model for analysis of ecological effects that may result from the introduction of transgenic organisms into the marine environment.

Chapter 4 focuses on the implications for risk assessment of uncontained trials of transgenic marine microorganisms. The chapter includes an assessment of the results of applications of terrestrial genetically engineered microorganisms (GEMs) and an examination of additional concerns that risk assessors may face when evaluating GEMs for use in the marine environment. The author considers the wide distribution of microorganisms in the oceans, the difficulty of ensuring containment of introduced species, the variety of reproductive and dispersal mechanisms available to marine microorganisms, and the problems that may arise if established GEM populations promote the flow of transgenes through marine ecosystems.

Chapter 5 examines legislation that may apply to uncontained applications of marine biotechnology products, including federal, regional, and state regulations that address the coastal environment, wildlife, and marine aquaculture. The author highlights areas in which statutory oversight may not be adequate and recommends actions to address weaknesses in the regulatory framework.

Chapter 6 considers the potential economic impact on government, industry, and society of introduced marine GEOs. Applying the methodologies of neoclassical, environmental, and ecological economics, the authors propose the use of case-specific cost-benefit analyses to assess the economic implications of transgenic marine organisms. They offer examples from the biotechnology, fishing, and tourism industries to illustrate potential economic risks and benefits associated with marine GEOs.

In Chapter 7, the authors summarize their findings and make recommendations to encourage an appropriate balance between environmental safety and economic development in the emerging field of marine biotechnology.

REFERENCES

Wommack, K.E., R.T. Hill, M. Kessel, E. Russek-Cohen and R.R, Colwell. 1992. Distribution of viruses in the Chesapeake Bay. Applied Environmental Microbiology 58(9):2965-2970.

Zilinskas, R.A., R.R. Colwell, D.W. Lipton and R.T. Hill. 1995. The Global Challenge of Marine Biotechnology: A Status Report on the United States, Japan, Australia and Norway. College Park, MD: Sea Grant College.

LIST OF CONTRIBUTORS

Balint, Peter J.
Center for Public Issues in Biotechnology
University of Maryland Biotechnology Institute
4321 Hartwick Road, Suite 500
College Park, MD 20740
School of Public Affairs
University of Maryland
College Park, MD 20740
Phone: (301) 403-8383
FAX: (301) 454-8123
E-mail: balint@wam.umd.edu

Colwell, Rita R., Ph.D., D.Sc.
President
University of Maryland Biotechnology Institute
4321 Hartwick Road, Room 550
College Park, MD 20740
Phone: (301) 403-0501
FAX: (301) 454-8123
E-mail: colwell@umbi.umd.edu

Gutrich, John J.
Environmental Science
The Ohio State University
2021 Coffey Road
Columbus, OH 43210
Phone: (614) 292-9329
FAX: (614) 292-0078
E-mail: gutrich.1@osu.edu

Hite, Diane, Ph.D.
Department of Economics and Center for Human Resource Research
The Ohio State University
403 Arps Hall
1945 North High Street
Columbus, OH 43210
Phone: (614) 292-5599
FAX: (614) 292-3906
E-mail: hite.10@osu.edu

Levin, Morris, Ph.D.
Center for Public Issues in Biotechnology
University of Maryland Biotechnology Institute
701 East Pratt Street, Suite 236, Columbus Center, Room 5-49
Baltimore, MD 21202
Phone: (410) 234-8880
FAX: (410) 234-8896
E-mail: levin@umbi.umd.edu

Stenquist, Susan
This work was completed while the author was with:
Center for Public Issues in Biotechnology
University of Maryland Biotechnology Institute
4321 Hartwick Road, Suite 500
College Park, MD 20740
Phone: (301) 403-8383
FAX: (301) 454-8123
Present affiliation:
Wildlife Conservation Society
185th Street and Southern Boulevard
Bronx, NY 10460
E-mail: sstenquist@mail1.wcs.org

Whiteman, Howard H., Ph.D.
Department of Biological Sciences
Murray State University
Murray, KY 42071
Center for Public Issues in Biotechnology
University of Maryland Biotechnology Institute
College Park, MD 20740
Phone: (502) 762-6753
FAX: (502) 762-2788
E-mail: howard.whiteman@murraystate.edu

Zilinskas, Raymond A., Ph.D.
Center for Public Issues in Biotechnology
University of Maryland Biotechnology Institute
4321 Hartwick Road, Suite 500
College Park, MD 20740
Phone: (301) 403-8383
FAX: (301) 454-8123
E-mail: zilinska@umbi.umd.edu

ACKNOWLEDGMENTS

We wish to acknowledge those whose assistance, cooperation, and support made this volume possible. We are indebted to the National Sea Grant College Program, Maryland Sea Grant College, and the University of Maryland Biotechnology Institute for financial support.

The following individuals reviewed drafts of the chapters and provided commentary and suggestions for improvement: Ronald M. Atlas, University of Louisville; Meryl Broussard, United States Department of Agriculture; Charles L. Brown, United States Department of Agriculture; James T. Carlton, Williams College; Mary Colligan, National Marine Fisheries Service; Daniel Gutrich, Jet Propulsion Laboratory, NASA; Anwarul Huq, University of Maryland; Chris Mantzaris, National Marine Fisheries Service; Richard McLaughlin, Mississippi-Alabama Sea Grant Legal Program; Gwendolyn McClung, United States Environmental Protection Agency; Douglas Parker, University of Maryland; Dennis A. Powers, Hopkins Marine Station; Roger Prince, Exxon Research and Engineering; Philip Sayre, United States Environmental Protection Agency; and Sivramiah Shantharam, United States Department of Agriculture. All remaining errors are our own.

We appreciate the time spent by the following persons who shared their invaluable expertise: Thomas Bell, United States Food and Drug Administration; Suzanne Bogren, University of Dayton; Elliot Entis, A/F Protein; Fred Hitzhusen, Ohio State University; John Matheson, United States Food and Drug Administration; Bradley Powers, Maryland Department of Agriculture; Robert Peoples, United States Fish and Wildlife Service; William Jay Troxel, United States Fish and Wildlife Service, and the state representatives who answered our questionnaire (see Appendix, Chapter 5).

In addition we wish to express our gratitude for the continuing encouragement offered by Francis Gutrich, Joyce Gutrich, Nancy Buschhaus, and Gail McKiernan, among many others.

Chapter 1

Risk Assessment for Uncontained Applications of Genetically Engineered Organisms

MORRIS LEVIN
University of Maryland Biotechnology Institute

1. INTRODUCTION

By applying the tools of modern molecular biology it is possible to produce genetically engineered organisms (GEOs) tailored to particular uses. During development, all field trials of transgenic products involve the potential for an accidental release into the open environment (Witt, 1990). The inherent danger to ecosystems must be assessed and managed through risk assessment protocols. Over the past decade, the use of agricultural products modified through biotechnology has moved beyond the testing stage, and many crops grown from genetically altered seeds are widely marketed. Formal procedures for assessing the associated risks are well established; to date, there have been no known adverse environmental effects from the release of terrestrial GEOs (Medley, 1996).

As research in marine biotechnology creates products potentially useful to the aquaculture and ocean bioremediation industries, the risk assessment procedures that have proved successful with land-based GEOs must be re-evaluated to determine their applicability to the marine environment. As is discussed in greater detail in Chapters 2, 3, and 4, the characteristics of the ocean that distinguish it from the terrestrial environment, particularly regarding expanded potential for reproduction and dispersal of introduced species, will present unprecedented challenges for risk managers. These concerns are compounded by gaps in the science base relating to marine ecology that make the potential impacts of marine GEO introductions less predictable than those resulting from terrestrial applications.

In general, the procedure for assessing and managing risk is analogous to a filter system designed to purify water. Each additional layer of filtration material increases water purity but also reduces output. Too few layers allow unwanted contaminants to pass; too many layers result in cessation of water flow. A balance must be struck between purity and supply. The role of risk assessment and risk management in the field of biotechnology is to ensure safety without preventing the development of useful products (i.e., to encourage both supply and purity).

The questions that need to be asked to evaluate ecological risk are fundamentally the same for any introduction of nonindigenous organisms into a new environment. The challenge for regulators is to obtain enough data to answer these questions with confidence. As has been the case in terrestrial field trials, the goal with regard to marine transgenics will be to balance concerns about potential adverse environmental and public health effects against the potential benefits of responsible scientific and commercial applications of this new technology.

This chapter (1) discusses the history of risk assessment and regulatory authority in the field of biotechnology; (2) considers risk assessment protocols already accepted for use with land-based agricultural applications; (3) evaluates the efficacy of applying existing protocols to open environment trials of genetically engineered marine organisms; and (4) makes recommendations to policy makers that, if implemented, would ensure appropriate risk management in the emerging field of marine biotechnology.

As emphasized in the Preface, the primary focus of this book is on the ecological and economic effects of developments in marine biotechnology and their implications for regulatory risk assessment. In this chapter, therefore, I give primary consideration to risk assessment procedures as they are used to evaluate potential adverse ecological effects; I do not attempt a full discussion of human health issues.

2. HISTORY OF REGULATION AND RISK ASSESSMENT OF GEOS

In 1986, the U.S. Office of Science and Technology Policy (OSTP) published its Coordinated Framework for the Regulation of Biotechnology (OSTP, 1985), which stated that existing statutes were sufficiently broad to regulate biotechnology products. This finding was based on the analysis conducted by the Biotechnology Science Coordinating Committee (BSCC) with input from the Environmental Protection Agency (USEPA), the Department of Agriculture (USDA), the National Institutes of Health (NIH), and the Food and Drug Administration (FDA). Though initially controversial, the finding that existing laws were adequate to provide oversight of recombinant DNA (rDNA) technology became the basis of regulatory policy in this area.

The determination that there was no need to seek new legislation was met with skepticism by parts of the scientific community, various public interest groups, and some members of Congress. In particular, there was concern that those who would have responsibility for risk assessment under this regime—the Recombinant DNA Advisory Committee (RAC) of NIH, for example—would be knowledgeable in the field of molecular biology but would not be as experienced in biology, microbiology, ecology, or other life sciences related to uncontained environmental trials. In response to this concern, RAC agreed to bring in representatives from relevant fields to broaden the risk assessment expertise. A decade of experience of field applications without apparent adverse effects gives considerable confidence in the processes in place. Although there are still those who advocate omnibus legislation regulating GEOs, Congress has thus far declined to take such action.

BSCC reviewed 107 existing laws, regulations, and guidelines and determined which acts and agencies would have jurisdiction over genetically engineered products (see Appendix III, Table 1). However, the two areas most relevant to the present inquiry, aquaculture and bioremediation, were not included in the discussions. EPA was designated as the agency responsible for safeguarding public health and the environment, USDA was given oversight authority over agricultural applications, and FDA was named the lead agency for food safety and animal drug issues. (As discussed in more detail below, it is this provision that has given FDA primary authority over the development of transgenic fish for marine aquaculture.)

FDA reviewed its authority under the Food, Drug and Cosmetic Act and found that it could assume its component of the responsibility for assessing the effects of biotechnology products without having to change its regulations. USDA and EPA, on the other hand, elected to develop new guidelines for uncontained applications of GEOs. All three agencies focused primarily on terrestrial products. No oversight policies were developed specifically for aquatic products at that time, though the USDA Agricultural Biotechnology Research Advisory Committee (ABRAC, now defunct) sponsored workshops on issues relating to transgenic fish research (ABRAC, 1992) and ultimately drafted a set of risk management guidelines (ABRAC, 1995). These latter documents, though having no formal standing, are presently being used by FDA assessors in the development of oversight procedures for transgenic fish and shellfish to be used in marine aquaculture (Matheson pers. comm., 1997).

Detailed descriptions of USDA and EPA organizational structure and policies can be found in Appendices I and II attached to this chapter. Further analysis of applicable statutes at the state and federal level can be found in Chapter 5.

3. RISK ASSESSMENT PROTOCOLS IN TERRESTRIAL TRIALS

The first field trial of a transgenic organism was of a bacterium designed to protect plants against frost. This test conducted in 1986, which successfully demonstrated the efficacy of the product and produced no adverse environmental effects (Levin and Strauss, 1993), has been followed by over 2000 plant and microorganism field applications in the terrestrial environment (GeneExchange 1994a, 1994b; Beck and Ulrich, 1993; PIP, 1993; Chasseray and Duesing, 1993; Levin, 1996). The fact that these trials have gone forward over the course of a decade with no known negative effects suggests that the risk assessment procedures in place have been successful in filtering out unsafe uses of GEOs (Medley, 1966). In this section I examine these procedures in some detail to provide background for a subsequent consideration of their applicability to uncontained trials of marine transgenics.

3.1 Risk Assessment Procedures

The risk assessment and risk management process, as it has evolved over the past decade to regulate GEOs, is complex and often idiosyncratic; that is, procedures vary on a case-by-case basis depending on the agency charged with oversight, the mandates of the relevant statute, and the specifics of the proposed trial. The unifying factor for risk assessors is to ensure consideration of the relevant items in "Points to Consider," a document developed by an interagency committee charged with identifying requirements for risk assessment of biotechnology products. EPA published its version of Points to Consider in 1990 (USEPA, 1990) and provided an updated version in June 1997 to meet requirements in the recent Final Rule for Microbial Biotechnology Products (USEPA, 1997a).

Although the basic questions that need to be asked remain the same, the Final Rule (USEPA, 1997a) differs from earlier USEPA regulations by establishing alternate procedures requiring different levels of information from the applicant depending on the nature of the organism to be tested. The establishment of three levels of application (Appendix I), each with differing information requirements, reflects a desire to minimize impact on applicants while maximizing protection of public health and the environment. The Final Rule clearly states that it covers only activities with commercial intent (under a broadened definition of what is commercial) and is based on an evaluation of the "newness" of a product. Newness is determined by considering whether the extent of the genetic differences between the GEO and the parent species crosses a threshold known as the intergeneric cutoff. This cutoff, defined in the Federal Register (FR) (USEPA, 1997a), is based on taxonomic distinctions. Despite these revisions, the basic requirement for an information set fulfilling the needs of a risk assessor based on the Points to Consider has not changed.

For USDA, key elements equivalent to those in the EPA Points to Consider document are contained in the agency's Application Form (USDA, 1990). (See Appendix III, Table 2, for a composite of the points included in the EPA and USDA documents.) The FDA, as noted above, has elected to employ the more detailed and extensive ABRAC documents as a basis for developing protocols to assess risks associated with biotechnology products for marine aquaculture.

In any event, to meet the requirements explicit in Points to Consider, applicants and agencies must follow several steps before an open field trial or product registration is approved (Appendix III, Table 3). First, the developer must present data describing the test organism and its potential ecological effects. Since authority for risk assessment is based on the empowering legislation, each agency will emphasize criteria specific to its area of oversight. Because responsibility for evaluating ecological risk has not been centralized in one agency, that task will fall to the regulators charged with oversight of the product in question. For example, risk assessors at the Center for Veterinary Medicine (CVM), a branch of FDA, are evaluating the potential ecological risks posed by transgenic salmon through their authority over animal drugs. The rationale is that since biotechnology products are not specifically targeted by statute, a growth hormone that an animal has been induced to produce internally through insertion of a foreign gene is the same, for regulatory purposes, as a growth hormone administered through conventional means. Therefore fish with transgenes that express growth hormone production fall under the authority of CVM.

No matter which agency is conducting the evaluation, all ecological risk assessment efforts require determination of two critical aspects of the proposed trial: the exposure anticipated and the hazard involved. Evaluating exposure involves determining what other organisms will be present at the test site and in what density. Evaluating hazard requires knowledge of how indigenous organisms, and other aspects of the local ecology, are likely to be affected by the test organisms.

Evaluating levels of hazard and exposure requires risk assessors to consider potential adverse effects. The level of scrutiny will depend upon how well known the GEO is and the extent to which it has been genetically manipulated. The National Academy of Sciences (NAS, 1989) and an independent group of scientists (Tiedje et al., 1989) have characterized this criterion in terms of the organism's "familiarity." With a product such as a well-studied bacterium engineered to contain a marker gene, this could mean simply asking if there are other closely related organisms for which it could be mistaken. For example, endowing a bacterium with the ability to utilize lactose to permit detection on a specific medium could result in misidentification of that organism by persons using standard taxonomic tests based on this property.

In contrast, organisms that have been genetically engineered to affect phenotypic expression are subject to more exhaustive review. Critical issues in determining risk in such full-scale reviews may include taxonomic position and life history. One example of the importance of taxonomy is that the more closely the test GEO is related to indigenous organisms the greater the likelihood of crossbreeding. In the terrestrial environment this means that introduced transgenic plants must be evaluated for their ability to cross-pollinate with native species. In the marine context, the ability of the test subject to interbreed and transmit the altered gene must be assessed.

As noted above, there is no standard procedure among agencies for conducting risk assessments. Each agency, and each department within each agency, organizes its risk assessment personnel in a different manner. The time required to approve or reject an application also varies within limits set by the applicable statute. For example, within EPA the office that conducts risk assessment under the Toxic Substances Control Act (TSCA) routinely forms committees to assess levels of exposure and hazard associated with each product to be evaluated. The committees analyze relevant factors including, for example, potential human health and ecological effects, exposure that may result during manufacturing, and the taxonomy of the organism. In addition, the field release protocol is reviewed. After the committees complete their deliberations, they hold a joint meeting wherein each presents its findings and a combined report is generated. On the other hand, the EPA office operating under authority from the Federal Insecticide, Fungicide, and Rodenticide Act (FIFRA) assigns a trial application to an assessor who parcels out the issues to individual specialists (selected for the occasion based on expertise in the subject at hand) and later combines the findings in a final report. Both TSCA and FIFRA risk assessors call upon outside experts to aid in evaluation, although the FIFRA office does so more extensively. FIFRA convenes Science Advisory Panels to examine applications while TSCA convenes Biotechnology Science Advisory Committees as needed. Also, at any point in the process, the agency may request further information from the applicant. USDA assessment procedures are similar to those of EPA under FIFRA. At FDA, assessors at CVM may call upon a fixed-membership advisory panel (Matheson pers. comm., 1997). This option is less flexible than the outside review available to assessors at EPA working under authority granted by TSCA or FIFRA.

In all cases, however, after exposure and hazard have been characterized by risk assessors, the results are passed on to a risk manager for final review and disposition. These findings provide a basis for the manager to reach a decision. The manager may (1) deny the application; (2) request additional data; (3) require a change in the product; (4) approve the application with special conditions attached; or (5) approve the application as presented. Examples of special conditions that may be imposed include confining the trial to a limited area, requiring protective physical constraints, or establishing strict monitoring and mitigation procedures. In the terrestrial environment, physical containment may include berms to catch run-off, fences to restrict entry, and bags over flowers to trap pollen. To ensure worker protection, the use of such items as hoods and respirators may be required. If the product in question is a pesticide, use may be limited to particular seasons (to protect migratory birds, for example).

To return to the filter analogy, the risk manager, working from reports generated by risk assessors, is responsible for approving projects that do not pose undue risk, while being careful to maintain appropriate environmental safeguards. Requirements that are too stringent result in a cumbersome, onerous process that discourages product development. Controls that are too lax result in an unacceptably high probability of adverse effects.

3.2 Existing Protocols

The EPA Points to Consider and the elements covered in the USDA Application Form are fundamentally the same; both were developed to ensure that all pertinent risk assessment information is available to the reviewer (Appendix III, Table 2). The ABRAC (1995) documents also cover the same questions, but require stricter adherence to a series of highly detailed flowcharts that offer less flexibility to regulators.

A comparison of general biotechnology risk assessment protocols promulgated by other nations and international organizations with those set out in the Points to Consider finds substantial agreement. For example, the scope of information an applicant for a permit must present as a basis for review is remarkably similar. A recent survey of nations in the Organization for Economic Cooperation and Development (OECD, 1995) demonstrated a 90% agreement among their points to consider. One major difference is that Canada and Australia place more emphasis on requirements regarding the degree of familiarity with the host organism than do other nations. Canada also requires review of naturally occurring microorganisms.

The basic approach inherent in these protocols correlates with suggestions put forth by the U.S. National Academy of Sciences (NAS) and the Ecological Society of America (Tiedje et al., 1989). In their view, there is a risk continuum. On one end of this continuum are easily contained, well understood host organisms such as cows with introduced bovine growth hormone. At the high risk end are less well known, less easily controllable organisms such as microbes engineered for bioremediation.

An important consideration is that risk assessment protocols are generally designed for small-scale experimental trials. Requirements for assessing the risks involved in large-scale commercial uses of transgenic organisms are less well defined. It cannot be assumed that data requirements for field tests will suffice for commercial releases, or that results from small-scale trials can be extrapolated to aid in understanding the risks involved in large-scale activities (Burke et al., 1994; Snow, 1997). For example, if a food crop genetically engineered to express production of a bacterial toxin that serves as a pesticide is planted on a commercial-scale, there is a much greater risk that insects will develop resistance than would be present in a small-scale trial, since the time scale and the size of the population exposed would be much larger.

A recent international symposium considered issues associated with large-scale uses of transgenic plants (Burke et al., 1994). Participants discussed the data that should be required and the risk assessment methodologies that would be most appropriate to insure credible evaluation of commercial releases. Minimizing potential impact on biodiversity, biogeochemical cycles, community structure, and other ecological factors was a primary focus. Although it was generally felt that engineered plants, because of their physical characteristics (nonmotile, controllable), need not be considered major hazards, the necessity of developing specific tests to identify such potential adverse effects as invasiveness and persistence was noted.

Crawley et al. (1993) have described an experimental approach that can be used to demonstrate the lack of invasiveness of a particular plant relative to the parent plant. This method of collecting and analyzing data allows risk assessment decisions to be based on actual data on a case-by-case basis. Generalization may be possible across closely related species and similar locations.

To conclude this section, it must be noted that the safety record is excellent for terrestrial applications of biotechnology products. There have been many releases over more than a decade, and none has resulted in observed adverse effects. This rate of success has been achieved without legislation specifically targeting GEOs and with each agency using its own protocols and organizational structure. This positive outcome can best be attributed to three factors: a bias toward conservative decision making on the part of risk assessors and managers in all agencies; the comparatively high degree of information available about terrestrial products and ecology; and the relative safety of biotechnology products compared, for example, to toxic substances and other environmental contaminants. In the following section we examine the appropriateness of applying the same procedures to marine biotechnology products.

4. RISK ASSESSMENT AND MARINE BIOTECHNOLOGY

Existing regulatory procedures for overseeing biotechnology products evolved in response to terrestrial applications. Knibb (1997) postulates that, based on the history of numerous releases of fish with laboratory induced changes (resulting from selective breeding, not genetic engineering), there is a negligible probability of ecological impact from transgenic fish that may escape containment. We must recognize, however, that commercial-scale utilization of genetically engineered products (terrestrial or marine/aquatic) does not have a long history and subtle or slow-to-emerge adverse effects may yet be detected. In addition, it must be pointed out that in most cases trials are not designed with ecological safety issues as the primary focus because in the initial stages of such trials strict containment is required (Snow, 1997; Levin, 1996; Burke et al., 1994). Therefore, several factors must be considered before applying terrestrial risk assessment procedures to marine trials of transgenic organisms.

First, gaps in the science base relating to marine/aquatic ecology must be examined to see if sufficient information for credible risk assessment is available. Historical case studies of exotic invasions can provide useful information, but a basic understanding of the biotic and abiotic characteristics of the oceans is required. As discussed in Chapters 2 through 4, marine ecosystems are less well understood than terrestrial systems and the ocean medium provides greater opportunities for reproduction and dispersal of introduced organisms. As a result, the potential environmental effects of GEO releases will be less predictable and more difficult to monitor and contain in the ocean environment.

Second, risk assessment protocols, particularly the Points to Consider, must be interpreted broadly enough to cover the added challenges presented by the marine environment. All the questions posed in the protocols that address terrestrial uses of GEOs apply to marine trials. However, since marine ecosystems are less well understood, applications submitted by industry for uncontained introductions of marine GEOs must be evaluated on a case-by-case basis and in a conservative manner that incorporates consultations with outside experts in marine biology and marine ecology. Development of marine microcosms that could be used to observe the effects of altered micro- and macroorganisms on their surroundings would provide basic data for risk assessors resulting in a more credible assessment capability.

4.1 The Science Base

Research and development in marine biotechnology has produced at least 14 species of fish and several species of mollusks and crustaceans to which foreign genetic material has been added (Hallerman and Kapuscinski, 1992a, 1990; NAL, 1993; Funkenstein et al., 1991; Chen, 1992) (Appendix III, Table 4). Many of these organisms are being engineered for more rapid growth and higher yield. To date, work in aquatic biotechnology has been limited to research in the laboratory and in ponds subject to environmental quarantine. Because the majority of experiments have occurred in conditions subject to strict containment, they offer little information that might directly apply to questions about safety in the open marine environment.

Appraising the effects of past dispersals and invasions of exotic organisms into new environments, however, can provide useful risk assessment information. For example, there have been many deliberate introductions of nonindigenous species for various purposes (Howarth, 1991). Data generated by these introductions can be analyzed to glean insights into the influence of new species on existing populations. Howarth (1991) describes case studies of the introduction of insects (e.g., *Levuana iridescens*), snails (e.g., *Gonaxis kibweziensis*), fish (e.g., *Gambusia* spp.), and plants (e.g., *Tibulus terrestris*). Among the critical characteristics he identifies that limit or enhance environmental effects are persistence, habitat range, phenotypic plasticity, host availability in the case of parasites, and behavior. Although these factors are drawn from analysis of nontransgenic introductions, Howarth (1991) notes that they are almost identical to those essential for assessment of the risks associated with the release of genetically engineered microbes, plants, and fish. (Ecological risk factors that apply in the marine environment are discussed in greater detail in Chapters 2, 3, and 4.)

Other examples of introductions, whether intentional or unintentional, and whether the organism persisted or suffered local extinction, can be used to assess the potential effects of introducing GEOs into the marine environment (Iwama et al., 1992). Global seagoing commercial activities, for example, often carry species outside their native habitats. Generally, organisms are adapted to their environment and will not persist in another setting. Occasionally a species transported to a foreign site will find an ecological niche in which it can establish itself (Chapter 3). In most terrestrial cases no effects have been noted; in a few cases beneficial effects have occurred; in a very few cases, adverse effects have been observed (Simberloff, 1985; Sharples, 1991). In his survey of biological invasions, Holdgate (1986) notes that data support the pattern that 10% of exotic species introductions may succeed, and of these only 10% may produce adverse effects. However, invasive exotic species that do establish themselves, though rare, can cause dramatic ecological disturbance. Well known examples include kudzu in the southeastern United States and the brown snake in Guam (Howarth, 1991). In contrast, the introduction of the Vedalia beetle in California is an example of a successful, beneficial introduction of an exotic species. In this case, crop damage caused by indigenous herbivorous pests was significantly reduced.

Increasing awareness of the applicability of information on dispersal of exotics to issues relating to the introduction of transgenics has raised interest in historical data. The natural bias towards studying and reporting examples of spectacular negative outcomes may lead to an overestimation of the potential for adverse effects posed by nonindigenous organisms. Studies of introductions of exotic species that have had neutral or positive outcomes are less well represented in the literature (Simberloff, 1985).

A conference on the subject sponsored by the International Committee for the Exploration of the Seas (ICES) in 1991 resulted in the publication of data based on relevant case studies (Aoki, 1991; Brock, 1992; Carlton, 1992; Zilinskas and Lundin, 1993). These evaluations of previous experiences with nonindigenous species reinforced the conclusion that in the majority of cases no adverse effects are noted.

Nevertheless, it is important to consider cases in which there have been negative effects to help build a credible base of experience for application of risk assessment protocols to the ocean environment. Four examples of the deleterious impacts of marine/aquatic exotics are: *Sargassum muticum* along the English and French coasts, *Hydrilla* in the Chesapeake Bay area, *Penaeus vannamei* in the Pacific Rim area, and zebra mussels in the Great

Lakes region. To illustrate one example, zebra mussels colonize intake and outlet pipes used by power plants and other installations, thereby reducing water flow through the ducts and interfering with industrial activity.

The transferals mentioned above were accidental, typically the result of stowaway exotics ejected into a new locale as ships cleared their bilge water. There have also been deliberate introductions of marine species into new sites for various purposes. For example, the African Tilapia, the black tiger shrimp (*Penaeus monodon*), and the white shrimp (*Penaeus orientalis*) were taken to Asian and Latin American countries for commercial purposes. *Gambusia affinis* and *Lebistes reticulatus* were introduced into parts of the world where malaria is widespread to reduce mosquito populations (Zilinskas and Lundin, 1993). These deliberate introductions, for commercial and pest control purposes, were successful and achieved their goals without observed negative effects. There have been many similar experiences in the terrestrial environment. For example, up to 90% of commercial food crops grown in the United States today are either introduced or selectively-bred species.

There is a rapidly growing body of literature addressing the causes and effects of damaging introductions (Allen, 1992; Carlton, 1989; Carlton and Geller 1993; Williamson and Gribbin, 1991). Many studies have been conducted focusing specifically on marine plants (Neushul et al., 1992), mollusks (Cembella and Shumway, 1994), shrimp (Carlton and Geller, 1993), and finfish (Chen, 1992). In general, examination of these case studies indicates that the basic questions to be asked for risk assessment are the same whether one is assessing the potential impact of terrestrial or marine introductions. These questions, as enumerated in the Points to Consider document, coincide with the factors listed by Howarth (1991); host and habitat range, behavior, and phenotypic plasticity are cited by Howarth and included in the guidelines (Appendix I).

The gathering of information based on past experience needs to be coupled with continuing research. Areas requiring further study include the potential for gene exchange in marine bacteria (Britschgi and Giovannoni, 1991) and concerns relating to the ubiquity of viruses in marine waters (Wommack et al., 1992). Basic research on marine ecology should be accelerated and expanded.

Despite the analogy that is often drawn between introductions of exotic and genetically engineered organisms, differences exist that must be taken into account. The possible ecological impact of releasing GEOs into natural marine environments has been described and reviewed previously (Hallerman and Kapuscinski, 1992b; Zilinskas and Lundin, 1993) and is discussed in detail in Chapters 3 and 4. The nature and extent of the impact of GEO introductions will depend upon two factors: the effect of the inserted genes on phenotypic expression in the organism and the effect of the modified organism on the environment to which it is introduced.

Hallerman and Kapuscinski (1992b) developed a paradigm of three classes of phenotypic adjustments that may occur in the transgenic organism. They state that GEOs may manifest changes relative to the parent organism in physiology, behavior, and/or tolerance of environmental variations. For example, a genetically engineered fish may be larger than the parent organism. Feeding patterns and predator/prey interactions may be altered in such a way as to cause changes in community structure and ecological balance (see Chapter 3). Zilinskas and Lundin (1993) suggest a list of potential adverse effects that could result from these changes: (1) disruption of local fauna; (2) degradation of the gene pool through elimination of wild species; and (3) introduction of new diseases to the locale. All of these effects have clear parallels in the terrestrial environment.

Additionally, selecting risk assessment practices requires prior evaluation of the ability to eliminate the introduced organism from the environment. This judgment must be made on a case-by-case basis. As in the terrestrial environment, it may be relatively easy to contain, monitor, and ultimately eradicate introduced aquatic plants. In contrast, it may be impossible to contain or later eliminate microorganisms once they are introduced into the open marine environment. Various species of shellfish and finfish will fall along a continuum between these two extremes. Each case must be assessed individually.

The effects of introduced GEOs on the natural marine environment are complex and difficult to foresee. The interactions that can be expected to occur between introduced marine GEOs and native ecosystems are similar to those involved in risk assessment of terrestrial products (Howarth, 1992; Burke et al., 1994). However, because scientific understanding of marine ecosystems is less comprehensive than knowledge of terrestrial environments, these interactions will move along poorly understood pathways and lead to results that are less predictable until better data and testing procedures become available.

In order to develop risk assessment protocols that can be used with confidence in marine aquaculture, it is necessary to identify and understand the mechanisms by which transgenics may have ecological impact. Hallerman and Kapuscinski (1991) stress that the level of risks presented by marine GEOs can be expected to be an "exacerbation of those experienced with nontransgenic cultured fish." At the same time, Iwama et al. (1992) and Harache (1992) both point out that there is extreme variation in the reactions of released natural fish to new environments and of the environment to the presence of new organisms, indicating that transgenic and natural organisms should be examined with equal care before release.

The implication for risk assessment is twofold. First, basic research must continue. Second, as in terrestrial applications, regulatory authorities must call upon scientists with relevant expertise—in cases involving marine GEOs, this will include calling upon experts in all aspects of marine ecology.

4.2 A Comparison of Marine Risk Assessment Protocols

Standardized risk assessment protocols that have been proposed in the U.S. to estimate the nature and extent of the likely impact of marine GEOs upon native stocks and aquatic communities follow the logic of terrestrial risk assessment; that is, they reflect the need to minimize the potential for adverse environmental effects while permitting the development of useful products (USDA, 1991). The regulations used for terrestrial introductions, including the new Final Rule on Microbial Products of Biotechnology issued by EPA (USEPA, 1997a), may ultimately be used to assess proposed marine releases but were not developed for that purpose. Under this new revision, TSCA regulations will cover genetically engineered marine bacteria designed for bioremediation purposes, but, as indicated above, the regulatory status of transgenic macroorganisms remains poorly defined.

Outside the United States, other nations and international entities have established guidelines specifically targeting marine transgenics. In the United Kingdom, for example, the Department of the Environment (UKDOE) Advisory Committee on Release to the Environment (ACRE) recently updated its procedures (UKDOE/ACRE, 1997). The new version requires adherence to a guidance document (UKDOE/ACRE, 1993) and to UK requirements for worker safety (UKHSE, 1996). Denmark and ICES have also promulgated regulations for the oversight of field trials of marine GEOs.

An examination of the risk assessment protocols established by the UK, Denmark, and ICES (Appendix III, Table 5) reveals substantial similarities, but some differences, when compared with the USDA/EPA requirements. The most significant differences distinguishing the international protocols designed for marine GEOs from the Points to Consider used in the U.S. for evaluation of terrestrial trials can be found in the eight elements listed in Table 5, Section H (see Appendix III). As can be seen in this table, the marine protocols also vary among themselves with respect to these issues; that is, no one document requires compliance with all eight elements. For example, only ICES includes socioeconomic analysis, and only ICES mandates the ultimate destruction of introduced organisms. Both ICES and

Denmark stress the importance of releasing only nonbreeding organisms. In addition Denmark requires adherence to a list of nonreleasable organisms. The Danish protocol assumes containment is not possible and mandates that all possible effects (i.e., worst case scenarios) be evaluated prior to release. In contrast, ICES assumes that containment is possible and proposes to study the effect of introducing small lots of genetically engineered organisms into the environment.

Another area of difference reflects the continuing debate as to how much emphasis should be given to monitoring production as opposed to focusing on the characteristics of the finished product. The UK document requires information about the manufacturing process. On the other hand, the ICES protocol postulates that the characteristics of the released organism are the key factors in determining potential impact, not how the organism was produced.

In considering the issue of whether to monitor production or just the finished product, the analogy of the filter applies. For example, not requiring oversight of the manufacturing process, in the case of ICES, removes one safeguarding layer from the filtering membrane (resulting in the possibility of a product being contaminated by a pathogen). This places a greater burden on comprehensive examination of the final product if the same level of public health and environmental protection is to be maintained.

In the United States, risk assessment protocols that apply to marine fish are being developed for aquaculture purposes by the FDA. FDA's statutory authority causes this agency to focus primarily on animal drug and food safety concerns. Potential ecological effects are addressed as a secondary issue. Socioeconomic factors, such as the repercussions of large-scale marine aquaculture on local economies and communities are not considered by the FDA, and may fall under the aegis of the Departments of Commerce or Labor. At this point, socioeconomic effects are not part of the environmental risk assessment process in the United States. However, the recent report of the Presidential/Congressional Commission on Risk Assessment (1997) urges that stakeholder groups be placed at the center of the process. If implemented, this would result in a significant change in risk assessment procedure and in the information required of applicants. Input from various stakeholder groups at all points in the process will result in more questions being raised relating to both scientific and socioeconomic concerns.

Although marine organisms are not specifically addressed in U.S. protocols designed to regulate GEO introductions, the information requirements identified in the EPA Points to Consider document can be interpreted broadly enough to cover marine transgenics. This can best be seen by comparing the elements in Tables 2 and 5 (see Appendix III). Table

5, which lists the information requirements laid out in international protocols for marine risk assessment, contains the same general elements found in Table 2, which lists the terrestrial risk assessment requirements in the U.S.

Policies pertaining to the use of genetically engineered aquatic organisms are currently under review both in the U.S. and abroad. As noted above, UKDOE released its fisheries protocol in 1997. In the U.S., Hallerman and Kapuscinski (1991) have submitted a position statement on transgenic fish to be considered for adoption by the American Fisheries Society, which takes into consideration science and safety matters and recognizes the political issues relating to legal jurisdiction (see Chapter 5).

5. DISCUSSION AND RECOMMENDATIONS

5.1 Discussion

Risk assessment must be conducted prior to any introduction of nonindigenous organisms, whether exotic or transgenic. It is important to determine the probabilities of escape, persistence, and reproductive success and to consider the potential ecological effects that may occur if the organism establishes itself (Possee and Bishop, 1992). In order for a risk assessment to provide a useful foundation for risk management decisions that will be accepted by the public at large, it must be based on sound science and be completed within a framework that has its objectives clearly defined. Thus, in all cases, the decision-making criteria, and the supporting information required, must be defined in advance. The management options that follow the risk assessment are spelled out in the particular legislation driving the assessment.

All recent proposals stress that adequate risk assessment must be based on scientific knowledge supported by continuing research. However, most field trials of land based GEOs have been conducted by academic researchers seeking a better understanding of particular transgenic organisms or by companies attempting to demonstrate the commercial viability of new products. Very few trials have been specifically designed to evaluate risk. As a result, there are significant gaps in our understanding of how introduced species may affect complex marine (or terrestrial) ecosystems. On the other hand, there are substantial benefits to be gained from exploitation of the products of marine biotechnology. The need to find a balance that encourages development while maintaining safeguards has engendered a spirited debate.

At one end of the spectrum, the Community Nutrition Group (CNI), a public advocacy organization, recently urged a halt to all open environment trials of GEOs until further research is completed (CNI, 1995). They argue that gaps in the science base preclude credible risk assessment because we do not have sufficient information at hand to estimate hazard to marine ecosystems. However, satisfying their credibility requirements would add many layers to the risk assessment filter, restricting the flow of potentially beneficial scientific and commercial applications of this new technology. In fact, there can never be complete understanding of complex natural systems; risk assessments, of necessity, are conducted based on available data. The decade-long record in the U.S. of successful terrestrial risk assessment of GEOs demonstrates that if conservatively applied in a case-by-case manner, less than complete data can be sufficient.

At the other end of the spectrum, in 1991 the U.S. Biotechnology Working Group of the President's Council on Economic Competitiveness of OSTP, while recognizing the importance of continuing research on the potential for environmental and social impact, stressed the need for the nation to maintain its competitive advantage in this emerging field. The Council stated GEOs "shall not be subject to federal oversight" unless there is substantial evidence to indicate "unreasonable risks." Unreasonable risks were defined as the case in which "full social and environmental costs exceed the cost of governmental intervention to redress it." The logical problem with this position is that it is not possible to make a determination that no "unreasonable" risk exists without first doing a risk assessment. Again, the case-by-case approach allows regulators to distinguish between proposed trials that merit cursory review and those that require extensive evaluation (i.e., to filter out trivial cases).

In addition, if efforts to utilize transgenic fish are to be successful, the general public must be fully informed about associated benefits and risks and have confidence that regulatory authority will be exercised effectively (Hallerman and Kapuscinski, 1990; Zilinskas and Lundin, 1993; Kapuscinski and Hallerman, 1990). For example, initial public reaction to the use of GEOs in the open environment in the U.S., Germany, and Holland was negative and hostile. Plants were uprooted, demonstrations were organized, and picketers marched. With the passing of time, as releases became more common and no adverse effects occurred, public opinion moderated. Nevertheless, there is still considerable consumer resistance to transgenic crop plants (Zehendorf, 1994). In order for biotechnology products to be accepted, the public must have input and be convinced of the credibility of the risk assessment and regulatory oversight (USGAO, 1988). Risk assessments must take into consideration how the trial may be perceived by the public at large in order to respond with convincing arguments that the proposed use of transgenic organisms is safe, if in fact an objective and thorough assessment leads to that conclusion. As noted above, a recent report (Presidential/Congressional Commission on Risk Assessment, 1997) stressed that all stakeholders should be involved in the risk assessment process from the outset.

To conclude this discussion, I reiterate the contention that the general questions to be asked in assessing risk are the same no matter the type of introduction proposed. The issues that must be considered do not change whether the organism in question is terrestrial or marine/aquatic, non-native or transgenic. The basic concerns, impact on biological community structure, on the overall ecology of the area, and on human health, require responses to the same fundamental questions: How will the released organism interact with its environment? What are the hazards? How much exposure is likely?

What will differ for regulators are the specific scientific data requirements and the breadth of outside expertise to be called upon. Thus, although biogeochemical cycles may be involved with releases of certain bacteria (e.g., *Thiobacillus*) and the food chain may be involved for salmon engineered to grow larger and faster, the issues of competition, survival and containment, for example, are common to risk assessment for either product. These topics require similar databases for evaluation of the probability of occurrence of adverse effects. It must be determined whether variations in temperature, humidity, salinity, and other physical parameters will affect the released organisms in terms of survival and fecundity rates and the potential for gene transfer. In many cases, data for the parental organism is available, and data for the engineered organism may be inferred or developed in microcosms.

Whether the risk assessment protocols to be used were specifically designed for marine biotechnology products (e.g., the ABRAC documents), or adapted from those developed for terrestrial applications (e.g., Points to Consider), they can be construed broadly enough to cover marine introductions. The problem lies not in the form of the protocol, but rather in the fact that scientific knowledge in the field of marine ecology is less comprehensive than that available for terrestrial ecosystems.

5.2 Recommendations

Procedures to assess and manage risk relating to the release of genetically engineered organisms in terrestrial trials have been applied successfully over the course of the past decade with no noted adverse effects (Medley, 1996). The same level of success can be expected if marine products are subjected to similar scrutiny and the added complexity of the ocean environment is taken into account. As a caveat, it is important to remember that most terrestrial trials have generally been small in scale, thereby reducing risk, and that even though monitoring continues there may be long-term, subtle effects that have not yet manifested themselves.

To minimize risk to the marine environment, appropriate and credible risk assessment strategies in the emerging field of marine biotechnology must be built with the following five primary considerations as a framework. First, a single agency should be made responsible for regulating marine biotechnology, and that agency should adopt a consistent, standard, scientifically rigorous protocol. Second, if regulations designed for terrestrial applications are to be utilized for marine organisms, they must take into account the unique characteristics of the ocean environment. There are important differences between terrestrial and marine species in terms of rates of propagation and ease of dispersal; furthermore, ecological interactions are more complex in the open ocean environment (see Chapters 2, 3, and 4). Third, procedures must be conservative but balanced. That is, they must err on the side of safety without placing onerous constraints on either the applicant or the regulatory agency. Fourth, applications for permits must be considered on a case-by-case basis. The experience with terrestrial uses of GEOs has taught that care must be taken when generalizing across species and locations. Finally, the public must be involved. Risk management safeguards must be objective, credible, and clearly understood. Complete transparency and broad opportunity for public participation in the development and implementation of risk assessment protocols are essential if scientific and commercial applications involving marine/aquatic release of transgenic organisms are to meet with public acceptance.

Finally, as discussed above, there are gaps in our understanding of ecologically important processes in the marine environment (Britschgi and Giovannoni, 1991; Wommack et al., 1992). In particular, these researchers raise concerns relating to the ubiquity of viruses in marine waters. To fill in these gaps, basic research in marine ecology should be accelerated and broadened. A wise approach under the constraints of limited research budgets would be to survey industry to determine the likely near term sites and organisms for marine trials of GEOs. Then targeted research could be focused on these areas and organisms to increase the science base needed to support credible risk assessment.

APPENDIX I: EPA

TSCA: Under Office of Pesticides, Pollution and Toxic Substances (OPPTS)
  TSCA requires risk assessment of manufactured products not subject to other statutes. The process of producing a pesticide sometimes involves producing intermediary chemical

products. These intermediates may be subject to TSCA. It is important to stress that TSCA coverage does not include microbial pesticides except under special circumstances. However, products such as bioremediation mixtures and fertilizers are covered and engineered fish may be covered. For example, a pesticide that is a killed *Pseudomonad* containing *Bacillus thuringiensis* (Bt) toxin that was inserted by genetic manipulation had to comply with both FIFRA and TSCA because the process involved producing an intermediate which was subject to TSCA. OPPTS published a proposed set of regulations on Sept. 1, 1994 (USEPA, 1994a), which became final April 11, 1967 (USEPA, 1997b), and effective on June 2, 1997, containing the regulatory requirements for microorganisms subject to TSCA. The 1997 version specifies data requirements for notification, which, at some level, include all previous data requirements. New to the 1977 version of the regulations are the Microbial Commercial Activities Notification (MCAN), and TERA (TSCA Experimental Release Application). TERA applications are for exemptions for research and development activities subject to other federal agencies or programs and certain biofertilizer applications. Thus, the limitation of TSCA to commercial activities is still clearly in place..

Fish are not included in any TSCA regulatory guidelines except as species to be protected from adverse effects from products being reviewed. However, this does not preclude the coverage of applications involving release of fish. Plants involved in bioremediation (i.e., aquatic or terrestrial phytoremediation) may also be covered.

FIFRA: Under Office of Pesticide Programs (OPP)
USEPA pesticide jurisdiction includes "Any substance or mixture of substances intended for preventing, destroying repelling or mitigating any pest." This includes plant regulators, defoliants, and desiccants (FIFRA, Definition).

EPA regulates microbial pest control agents and has the authority to require data generation under FIFRA. The agency has prepared Registration Eligibility Documents (REDs) for some microorganisms, describing in detail the data requirements for registration. REDs are developed specifically for reregistration of products and reflect the agency's data requirements. The requirements for registration and experimental use are contained in U.S. Code of Federal Regulations (CFR) 40, Parts 158 and 172 respectively. These documents have been revised periodically to update requirements in response to advances in scientific methodology and knowledge since the first products of a particular type were registered. EPA's Subdivision M Pesticide Assessment Guidelines contain the protocols that must be followed to generate data to meet registration and Experimental Use Permit (EUP) requirements. There are no REDs for fish; however, the specific data describing the impact of a regulated product on fish is required, and specific fish and testing protocols are described in Subdivision M (USEPA, 1994b). Registration requirements are specified in the Federal Register (USEPA, 1988). All data are available to the public (USEPA, 1987).

Scientific Considerations for Points to Consider for Environmental Testing of Microorganisms (adapted from USEPA, 1990; 1997)
   I. Summary
     A. Present a summary of the proposed trial, including objectives and significance
   II. Genetic Considerations of Modified Organisms to be Tested
     A. Characteristics of the nonmodified parental organism
       1. Information on identification, taxonomy, source and strain
       2. Information on organism's reproductive cycle and capacity for genetic transfer
     B. Molecular biology of the transgenic organism
       1. Introduced genes

a. Source and function of the DNA sequence used to modify the organism to be tested in the environment
   b. Identification, taxonomy, source, and strain of organism donating the DNA
2. Construction of the modified organism
   a. Describe the method(s) by which the vector with insert(s) has been constructed include diagrams as appropriate)
   b. Describe the method of introduction of the vector carrying the insert into the organism to be modified and the procedure for selection of the modified organism
   c. Specify the amount and nature of any vector and/or donor DNA remaining in the modified organism
   d. Give the laboratory containment conditions specified by the NIH guidelines for the modified organism
3. Genetic Stability and Expression
   a. Present results and interpretation of preliminary tests designed to measure genetic stability and expression of the introduced DNA in the modified organism

III. Environmental Considerations
  A. The intent of gathering ecological information is to assess the effects on survival reproduction and/or dispersal of the modified organism. For this purpose, information should be provided where possible and appropriate on:
     1. relevant ecological characteristics of the nonmodified organism
     2. the corresponding characteristics of the modified organism
     3. the physiological and ecological role of donated genetic sequences in the donor and in the modified organism(s)
  B. For the following points, provide information where possible and appropriate on the nonmodified organism and a prediction of any change that may be elicited by the modification
     1. Habitat and geographical distribution
     2. Physical and chemical factors that can affect survival, reproduction and dispersal
  C. Biological interactions
     1. Host range
     2. Interactions with an effects on other organisms in the environment including effects on competitors, prey, hosts, symbionts, predators, parasites and pathogens
     3. Pathogenicity, infectivity, toxicity, virulence or as a carrier (vector) of pathogens
     4. Involvement in biogeochemical or in biological cycling processes (e.g., mineral cycling, cellulose and lignin degradation, nitrogen fixation, pesticide degradation)
     5. Frequency with which populations undergo shifts in important ecological characteristics such as those listed in lll. C. points I through 4 above
     6. Likelihood of exchange of genetic information between the modified organism and other organisms in nature

IV. Proposed Field Trials
  A. Prefield trial considerations
     1. Provide data related to any anticipated effects of the modified microorganism on target and nontarget organisms from microcosm, greenhouse and/or growth chamber experiments that simulate; trial conditions. The methods of detection and sensitivity of sampling techniques and periodicity of sampling should be indicated. These studies should include survival of the modified organism, replication of the modified organism dissemination of the modified organism by wind, water, soil mobile organisms and other means.

B. Conditions of the trial
  1. Describe the trial involving release of the modified organism into the environment
     a. Specify number(s) of organisms and methods of application
     b. Provide information (including diagrams) of the experimental location and the immediate surroundings
     c. Describe characteristics of the site that would influence containment or dispersal
  2. If the modified organism has a target organism, provide the following
     a. Identification and taxonomy
     b. The anticipated mechanism and result of the interaction between the released microorganism and the target organism
C. Containment/mitigation
  1. Research must be conducted or supervised a technically qualified individual
  2. Appropriate records must be maintained
  3. Specify and provide details (efficacy, disposal) of mitigation and emergency termination procedures

APPENDIX II: USDA

USDA: Courtesy Permit Requirements

Courtesy permits are issued to provide the applicant with assurance that the product is not a regulated article. The applicant must prepare a one to two paragraph letter stating that the proper document(s) (e.g., USDA List of Regulated Articles) has been examined or that the product is not covered for some other reason. USDA will respond with approval, and thus the applicant will have a document that can be used to assure state and local regulators that federal regulations have been complied with. In the case of disapproval, USDA will recommend either the notification or permit application procedure (USDA/APHIS, 1991).

Notification Requirements (Applies to Plants Only)

A transgenic plant is eligible for the notification process if it meets six requirements:

1. The plant is one of the following species: corn, soybean, potato, tobacco, cotton, tomato, or any additional species deemed safe by USDA (and meeting performance standards in parts 2-6, below).

2. The introduced material is "stably integrated" into the plant genome.

3. The function of the introduced material is known and it will not produce plant disease.

4. The introduced material does not cause infection or toxicity to nontarget organisms.

5. The introduced genetic material does not pose a significant risk of the creation of any new plant viruses (material must be noncoding or common in the area where the release will occur).

6. The plant conforms to performance standards, which describe shipping procedures, viability of the organism, test procedures, test termination mechanism, follow-up procedures to destroy "volunteers," with notification to USDA/APHIS of procedures. Such notification is due six months after termination of the trial. Applicants must include how the observations were made and provide an analysis of the significance. Notification shall take place 10 days in advance of interstate movement and 30 days in advance of interstate release or importation. BBEP will respond within the designated time. Denial results in the need to apply for a permit.

Note: The above is a summary of FR Mar 31, 1993 (USDA 1993a, 1993b). The FR should be consulted for exact wording and definitions.

Risk Assessment for Uncontained Applications of Genetically Engineered Organisms

Permit Requirements

The 1993 FR established 180 days as the time required for review of a permit application. This is an increase of 60 days from the prior time limit. The extension is due to a 15-day increase in the public comment time. The public now has a 60-day comment period.

Permits require the applicant to provide all information listed on the Permit Application form.

For regulated engineered items (those requiring permits), USDA requires:

1. A description of expression of the altered genetic material: how it differs from the parental material, and how the material is integrated in the transgenic product.

2. A detailed description of the molecular biology used to produce the product.

3. A detailed description of where the donor, recipient, and vector were collected, developed, and produced.

4. A detailed description of the trial and the production methodology.

5. A detailed description of containment procedures and procedures to prevent contamination of the donor, recipient, and vector constituents of the product and the product itself.

6. A detailed description of site and type of facility involved (e.g., greenhouse, growth chamber) and the uses and distribution of the product. This description is to include manufacturing location, sale and distribution information.

7. A detailed description of procedures to prevent escape and dissemination of the donor, the vector, and of the recipient organism.

8. A detailed description of methods for final disposition of the product.

APPENDIX III: TABLES

Table 1. Agencies Responsible for Approval of Commercial Biotechnology Products

Biotechnology Products	Responsible Agencies
Foods/food additives	FDA (a), FSIS (b)
Human drugs, medical devices, and biologics	FDA
Animal drugs	FDA
Animal biologics	APHIS
Other contained uses	EPA
Plants and animals	APHIS (d), FSIS (b), FDA (c)
Pesticide microorganism release	EPA (e), APHIS (d)
Other uses (microorganisms):	
Intergeneric combination	EPA (e), APHIS (d)
Intrageneric combination:	
Pathogenic source organism	
Agricultural use	APHIS
Nonagricultural use	EPA (e), APHIS (d) nonpathogenic source
Organisms	EPA report
Nonengineered pathogens	
Agricultural use	APHIS
Nonagricultural use	EPA (e), APHIS (d)
Nonengineered nonpathogens	EPA report

a: Designates lead agency where jurisdictions may overlap
b: FSIS (Food Safety and Inspection Service) is responsible for food use
c: FDA is involved when in relation to a food use
d: APHIS (Animal and Plant Health Inspection Service) is involved when the microorganism is a plant pest, animal pathogen, or related product requiring a permit
e: EPA requirement will only apply to environmental release under a "significant new use rule" that EPA intends to propose
Source: 51 FR 23339

Table 2. General Comparison: EPA-OPPT points to consider/USDA-APHIS Elements

Topic	EPA-OPPT considers (a)	USDA-APHIS requires (b)
A. Taxonomy		
Taxonomic position of recipient	Yes	Yes
Phenotype of recipient	Yes	Yes
Genotype of recipient	Yes	Yes
Characteristics of test organism	Yes	Yes
B. Construction		
Construct analysis	Yes	Yes
Final construct	Yes	No
Construction of test organism	Yes	Yes
C. Health and environment		
Phenotype of test organism	Yes	Yes
Toxicity testing	Yes	No
Human health effects	Yes	No/yes (c)
Possible environmental effects	Yes	No
Fate of organism/fate of inserted DNA	Yes/yes	Yes/yes
D. Manufacturing		
Site of manufacture	Yes	Yes
Production process	Yes	No
Containment during production	Yes	No
Production volume reported	Yes	No
Worker exposure at site of manufacture	Yes	No
E. Worker onsite exposure		
Onsite containment	Yes	Yes
Intended onsite volume release information	Yes	Yes
Worker exposure at test site	Yes	Yes
F. Release protocols		
Objectives/purpose	Yes	Yes
Site description	Yes	Yes
Design of experiment/field test	Yes	Yes
Onsite containment	Yes	Yes
Method of application	Yes	Yes
Termination		
(a) Normal conditions	Yes	Yes
(b) Emergency conditions	Yes	Yes
Mitigation methods	Yes	Yes
Monitoring	Yes	Yes
Sampling	Yes	No

Risk Assessment for Uncontained Applications of Genetically Engineered Organisms

Topic	EPA-OPPT considers (a)	USDA-APHIS requires (b)
Recordkeeping	Yes	No
G. Identification of applicant		
Principal investigator	Yes	Yes
Sponsoring organization	Yes	Yes
H. Origin of microorganisms and vectors		
Donor organism	Yes	Yes
Recipient organism	Yes	Yes
Vectors	Yes	Yes

a: USEPA Microbial products of Biotechnology, Final Regulation FR April 11 1997.
b: USDA User's Guide, USDA, 1991; and Permit Application Form (APHIS Form 2000), USDA, 1989
c: Soilborne Microorganisms National Biological Impact Assessment Program Permit Application System.

Table 3. Components of Risk Assessment

Steps to be followed
1. Request for review from developer
2. Preliminary evaluation by agency
3. Estimate of exposure and hazard
4. Characterization of risk (beneficial, adverse, or no effect)
5. Risk management decision and action
6. Selection of option based on specific statute (approve, reject, or add special conditions)

Table 4. Fish Species That Have Been Genetically Engineered

Fish (Variety)
Carp (Common, Indian Major, and Mud)
Catfish (Channel)
Goldfish
Loach
Medaka
Pike (Northern)
Salmon (Atlantic, Pacific, and Coho)
Sea Bream
Silver Crucian
Tilapia (Nile)
Trout (Rainbow)
Wuchang Fish
Zebra Fish

Table 5. Comparison of Marine Risk Assessment Protocols

Topic	Denmark (a)	UK (b)	USDA (c)	ICES (d)
A. Taxonomy				
Taxonomic position of recipient	Yes	Yes	Yes	Yes
Phenotype of recipient	Yes	Yes	Yes	Yes
Genotype of recipient	Yes	Yes	Yes	Yes
Characteristics of test organism	Yes	Yes	Yes	Yes
B. Construction				
Construct analysis	Yes	Yes	Yes	No
Final construct	Yes	Yes	Yes	No
Construction of test organism	Yes	Yes	Yes	No
C. Health and environment				
Phenotype of test organism	Yes	Yes	Yes	No
Toxicity testing	No	No	No	No
Human health effects	No	No	No	No
Possible environmental effects	Yes	Yes	Yes	Yes
Fate of organism/fate of inserted DNA	Yes	Yes	Yes	No
Determine access to unused or susceptible ecosystems	Yes	Yes	Yes	Yes
Introduce small lots and study effects	N/a	No	Yes	Yes
D. Manufacturing				
Site of manufacture	No	Yes	No	Yes
Production process	No	No	No	Yes
Containment during production	No	Yes	Yes	No
Production volume reported	No	No	No	Yes
Worker exposure at site of manufacture	No	No	No	No
E. Worker onsite exposure				
Onsite containment	No	Yes	Yes	N/a
Intended onsite volume release information	No	Yes	Yes	No
Worker exposure at test site	No	No	No	No
F. Release protocols (UK ACRE REQ.)				
Objectives/purpose	Yes	Yes	Yes	Yes
Site description	Yes	Yes	Yes	Yes
Design of experiment/field test	Yes	Yes	Yes	Yes
Onsite containment	N/a	Yes	Yes	Yes
Method of application	Yes	Yes	Yes	Yes
Termination	Yes	Yes	Yes	Yes
(a) Normal conditions	Yes	Yes	Yes	Yes
(b) Emergency conditions	Yes	Yes	Yes	Yes
Mitigation methods	Yes	Yes	Yes	Yes
Monitoring	Yes	Yes	Yes	Yes
Sampling	Yes	Yes	Yes	Yes
Recordkeeping	Yes	Yes	Yes	Yes
G. Identification of applicant				
Principal investigator	Yes	Yes	Yes	Yes
Sponsoring organization	Yes	Yes	Yes	Yes
H. Specific additions				
Test approval required				

Topic	Denmark (a)	UK (b)	USDA (c)	ICES (d)
(a) Institutional biosafety committee	N/a	No	Yes	N/a
(b) State and/or local agencies	N/a	No	Yes	N/a
(c) Federal agency or agencies	Yes	Yes	No	Yes
Destroy original organisms	No	No	No	Yes
Socioeconomic analysis	No	No	No	Yes
Extended post-release monitoring	No	No	No	Yes
Document disease history of released organism	Yes	No	No	No
Release nonbreeding organisms only	Yes	No	No	Yes
Create list of nonreleasable species	Yes	No	No	No
Create list of nonacceptable parasites, diseases	Yes	No	No	No

a: Denmark protocol for aquatic release (assumes containment not possible); need to establish reserves of potentially affected
species; need to estimate fishing pressure as a result of release.
b: UK guidelines: Department of Environment/ACRE. 1997. Guidance for Experimental Releases of Genetically Modified
Fishes and 1996 Guidance Note 1.
c: USDA/ARS
d: ICES recommendations for nontransgenic aquatic release: sterilize effluent from quarantine and brood locations.

REFERENCES

Agricultural Biotechnology Research Advisory Committee (ABRAC). 1995. Performance Standards for Safely Conducting Research with Genetically Modified Fish and Shellfish. Documents 95-04 and 95-05. Washington, DC: USDA.

Allen, S.K., Jr. 1992. Issues and opportunities with inbred and polyploid species. In: M.R. DeVoe (ed.). Introductions and Transfers of Marine Species: Achieving a Balance Between Economic Development and Resource Protection. Hilton Head, SC: South Carolina Sea Grant Symposium.

Aoki, T. 1991. The application of biotechnology in fish pathology in Japan. 55. Baltimore, MD: International Marine Biotechnology Conference.

Beck. C.I. and T.H. Ulrich. 1993. Environmental release permits. Biotechnology 11:1524-1529.

Britschgi, T.B. and S.J. Giovannoni. 1991. Phylogenic Analysis of a Natural Bacterioplankton Population by rRNA Gene Cloning and Sequencing. Applied Environmental Microbiology 57:1707-1713.

Brock, J.A. 1992. Procedural requirements for marine species introductions into and out of Hawaii. In: M.R. DeVoe (ed.). Introductions and Transfers of Marine Species: Achieving a Balance Between Economic Development and Resource Protection 51-54. Hilton Head, SC: South Carolina Sea Grant Symposium.

Burke, T., R. Seidler and H. Smith. Ecological implications of transgenic plants. Molecular Ecology 3(1):1-89.

Carlton, J.T. 1989. Man's role in changing the face of the ocean: Biological invasion and implications for conservation of near-shore environments. Conservation Biology 3(3):265-273.

Carlton, J.T. 1992. Marine species introductions by ship's ballast water: an overview. In: M.R. DeVoe (ed.). Introductions and Transfers of Marine Species: Achieving a Balance Between Economic Development and Resource Protection 23-26. Hilton Head, SC: South Carolina Sea Grant Symposium.

Carlton, J.T. and J.B. Geller. 1993. Ecological roulette: the global transport of nonindigenous marine organisms. Science 261:78-82.

Cembella, A. and S.E. Shumway. 1994. Sequestering and biotransformation of paralytic shellfish toxins in scallops: Food safety implications of harvesting wild stocks and aquaculture biotechnology. In: OECD Secretariat (ed.). The Proceedings of the Symposium on Aquatic Biotechnology and Food Safety 69-78. Paris: OECD.

Chasseray, E. and J. Duesing. 1993. Field trials of transgenic plants. AGRO Food Industry 1-10. Basel, Switzerland.

Chen, Z.L. 1992. Field releases of recombinant bacteria and transgenic plants in China. 53-54. Biosafety results of field tests of genetically modified plants and microorganisms. Biol. Bund. Land und Forst. Braunschweig, Germ.

Code of Federal Regulations (CFR) Title 40 part 152 (registration Procedures), part 158 (data requirements). Washington, DC: U.S. Government Printing Office.

Community Nutrition Institute (CNI). 1995. Transgenic Fish: The next threat to biodiversity. Washington, DC: CNI.

Crawley, M.J., R.S. Halls, M. Rees, D. Kohn and J. Buxton. 1993. Ecology of transgenic oilseed rape in natural habitats. Nature 363:620-623.

Funkenstein, B., B. Cavari, B. Moav, O. Harari and T.T. Chen. 1991. Gene transfer of growth hormone in the Gilthead Sea bream and characterization of its pregrowth hormone 93. Baltimore, MD: International Marine Biotechnology Conference.

GeneExchange. 1994a. Experimental releases of genetically engineered organisms. 4(4):12. Washington, DC: Union of Concerned Scientists.

GeneExchange. 1994b. Experimental releases of genetically engineered organisms. 5(2):12. Washington, DC: Union of Concerned Scientists.

Hallerman E.M. and A.R. Kapuscinski. 1990. Transgenic fish and public policy: regulatory concerns. Fish 15(1):12-20.

Hallerman, E.M. and A.R. Kapuscinski. 1991. Recent developments in public policies regulating the development of transgenic fishes. Baltimore, MD: International Marine Biotechnology Conference.

Hallerman, E.M. and A.R. Kapuscinski. 1992a. Ecological and regulatory uncertainties associated with transgenic fish. In: C. L. Hew and G.L. Fletcher (eds.). Transgenic-fish 209-228. Singapore: Singapore World-Scientific.

Hallerman E.M. and A.R. Kapuscinski. 1992b. Ecological implications of using transgenic fishes in aquaculture. ICC MSS 194 31-56.

Harache, Y. 1992. Pacific Salmon in Atlantic Waters. ICC MSS 31-55.

Holdgate, M.W. 1986. Survey and conclusions: characteristics and consequences of biological invasions. Philosophical Transactions of the Royal Society of London 314:733-742.

Howarth, G. 1991. Environmental impacts of classical biological control. Annual Review of Entomology 36:485-509.

Howarth, W. 1992. Regulating the introduction of freshwater fish: the UK, the EC and beyond. ICC MSS 21-30.

Iwama, G.K., J.C. McGreer and N.J. Bernier. 1992. The effects of stock and rearing history on the stress response in juvenile coho salmon. ICC MSS 67-84.

Kapuscinski A.R. and E.M. Hallerman. 1990. Transgenic fish and public policy: anticipating environmental impacts of transgenic fish. Fish 15(1):2-12.

Knibb, A. 1997. Risk from genetically engineered and modified marine fish. Trans. Res. 6:56-67.

Levin M.A. 1996. Risk Assessment. AAAS 1996.

Levin M.A. and H. Strauss. 1993. Overview of risk assessment. In: M.A. Levin and H. Strauss (eds.). Risk Assessment in Genetic Engineering 1-17. New York: McGraw Hill.

Matheson, J. 1997. Personal communication. (Matheson is with the Center for Veterinary Medicine at FDA in Rockville, MD.)

Medley, T. 1996. Evaluating risks of agricultural products. AAAS Feb. 1966.

National Academy of Sciences (NAS). 1989. Field Testing Genetically Modified Organisms: Framework for Decisions. Washington, DC: National Academy Press.

National Agricultural Library (NAL). 1993. Transgenic Fish Research: A Bibliography. USDA Number 117. Beltsville, MD: USDA.

Neushul, M., C.D. Amsler, D.C. Reed and R.J. Lewis. 1992. Introduction of marine plants for aqua culture purposes. In: A. Rosenfield R. Mann (eds.). Dispersal of Living Organisms into Aquatic Ecosystems 103-138. College Park, MD: Maryland Sea Grant College.

Office of Science and Technology Policy (OSTP). 1985. Coordinated framework for the regulation of biotechnology: establishment of the Biotechnology Science Coordinating Committee. FR 50: 47174-47195.

Organization for Economic Cooperation and Development (OECD). 1995. Analysis of information elements in assessment of certain products of modern biotechnology. OECD Environmental Monitor. 100.

PIP Newsletter. 1993. Field release of modified plants. Sept. 1993.

Possee R.D. and D.H.L. Bishop. 1992. Safety tests with genetically engineered baculovirus pesticides. 90-98. Biosafety results of field tests of genetically modified plants and microorganisms. Biol. Bund. Land und Forst. Braunschweig, Germ.

Presidential/Congressional Commission on Risk Assessment and Risk Management. 1997. A Public Health Approach to Environmental Protection. Washington, DC.

Sharples, F.E. 1991. Ecological Aspects of Hazard Identification. In: M.A. Levin and H. Strauss (eds.). Risk Assessment in Genetic Engineering. New York: McGraw Hill.

Simberloff, D. 1985. Predicting Ecological Effects of Novel Entities. In: H.O. Halvorsen, D. Pramer and M. Rogul (eds.). Engineered Organisms in the Environment. Washington, DC: ASM.

Tiedje, J.M., R.K. Colwell, Y.L. Grossman, R.E. Hodson, R.E. Lenski, R.N. Mack and P.J. Regal. 1989. The Planned Introduction of Genetically Engineered Organisms: Ecological Considerations and Recommendations. Ecology 70(2):298-315.

United Kingdom Department of the Environment (UKDOE) Advisory Committee on Release to the Environment (ACRE). 1993. Guidance Note 1. London: UKDOE/ACRE.

United Kingdom Department of the Environment (UKDOE) Advisory Committee on Release to the Environment (ACRE). 1997. Guidance for Experimental Releases of Genetically Modified Fish. London: UKDOE/ACRE.

United Kingdom Health and Safety Executive (UKHSE). 1996. Guide to Genetically Modified Organisms, Contained Use (as amended 1996). London: UKHSE.

United States Department of Agriculture (USDA). 1990. A User's Guide: Biotechnology Permits USDA/APHIS/BBED Hyattsville, MD: USDA.

United States Department of Agriculture (USDA). 1991. Agricultural Biotechnology Research Advisory Committee. Supplement to Minutes: Guidelines Recommended to USDA for Research Involving Planned Introduction of Genetically Modified Organisms. Washington, DC: Office of Agricultural Biotechnology, USDA.

United States Department of Agriculture (USDA). 1993a. Genetically Engineered Organisms and Products; Final Rule. Federal Register Mar. 31 1993 17044-17059.

United States Department of Agriculture (USDA). 1993b. Field Tests of Genetically Modified Organisms and Products. CSRS. Washington, DC: USDA.

United States Department of Agriculture (USDA) Animal and Plant Health Inspection Service (APHIS) Permits. 1991. User's Guide for Introducing Genetically Engineered Plants and Microorganisms. USDA Technical Bulletin 1783. Washington, DC: USDA.

United States Environmental Protection Agency (USEPA). 1987. Disclosure of Reviews of Pesticide Test Data. USEPA/OPTS. Federal Register 50 #229 18833-18835.

United States Environmental Protection Agency (USEPA). 1988. Pesticide Registration Procedures; Pesticide Data Requirements; Final Rule. 40 CFR parts 153, 156, 158, 162, and 163. FR 53 #88. 15952-15999.

United States Environmental Protection Agency (USEPA). 1990. Points To Consider in the Preparation and Submission of TSCA Premanufacture Notices for Microorganisms. Program Development Branch. Washington, DC: USEPA.

United States Environmental Protection Agency (USEPA). 1994a. Microbial Products of Biotechnology: Proposed Regulation Under the Toxic Substances Control Act: Proposed Rule. Federal Register 59 45526-45585.

United States Environmental Protection Agency (USEPA). 1994b. Microbial EUPs, Notifications and Transgenic Plant pesticides. Washington, DC: USEPA.

United States Environmental Protection Agency (USEPA). 1997a. Microbial Products of Biotechnology; Final Rule. Federal Register 62:2-45.

United States Environmental Protection Agency (USEPA). 1997b. Points to Consider in the Preparation of TSCA Biotechnology Submissions for Microorganisms. USEPA/OPPTS, Washington DC.

United States General Accounting Office (USGAO). 1988. Biotechnology: Managing the Risks of Field Testing Genetically Engineered Organisms. GAO/RCED-88-27. Washington, DC: USGAO.

Williamson, P. and J. Gribbin. 1991. How plankton change the climate. New Scientist 129:48-52.

Witt, S.C. 1990. Biotechnology, microbes and the environment. San Francisco, CA: Center for Science Information.

Wodwark, M., D. Penman and B. McAndrew. 1994. Genetically Modified Fish Populations. Institute of Aquaculture, University of Sterling, UK.

Wommack, K.E., R.T. Hill, M. Kessel, E. Russek-Cohen and R.R. Colwell. 1992. Distribution of viruses in the Chesapeake Bay. Applied Environmental Microbiology 58:2965-2970.

Zehendorf, B. 1994. What the public thinks about biotechnology. Biotechnology 12:870-873.

Zilinskas, R.A. and C.G. Lundin. 1993. Marine Biotechnology and Developing Countries. World Bank Discussion Paper 210. Washington, DC: World Bank.

Chapter 2

Characteristics of Marine Ecosystems Relevant to Uncontained Applications of Genetically Engineered Organisms

JOHN J. GUTRICH,[1] HOWARD H. WHITEMAN,[2] AND RAYMOND A. ZILINSKAS[3]
[1]*Ohio State University,* [2]*Murray State University, and* [3]*University of Maryland Biotechnology Institute*

1. INTRODUCTION

Striking differences exist between marine, freshwater, and terrestrial systems in terms of physical and biological processes. The physical parameters of ocean environments and the life history strategies of marine organisms are salient examples of these disparities. Certain features of marine systems make it difficult if not impossible to ensure strict containment of genetically engineered organisms (GEOs) introduced into marine environments and, therefore, have profound implications for risk assessment and management. For example, water movement alters the boundary layers around marine organisms, transports nutrients and waste products, provides the means for long distance dispersal, assists migrations, and influences predator/prey and parasite/host encounter rates. As a result, the physical and biological characteristics of the marine environment may increase the probability of an accidental introduction of transgenic organisms into natural ecosystems and create novel problems for assessing and minimizing negative effects in the event of such a release.

In this chapter, we examine in detail the distinctive attributes of the marine environment that are likely to complicate risk assessment and management. Among abiotic features, we focus on the spatial scale of the marine environment, the characteristics of water as a life-support medium, and the effects of temperature gradients and oceanic movements on ecosystem boundaries. Among biotic features, we consider marine life history strategies and the complexity of marine food webs. This chapter lays the groundwork for the analyses of potential ecological effects of introduced marine transgenics (both macro- and microorganisms) that are detailed in Chapters 3 and 4.

2. ABIOTIC FEATURES

Recognition of the complex interactions among physical, chemical, and biological processes in marine systems has led to a growing body of literature addressing components that emphasize this interplay (Hartline, 1980; Lewis et al., 1983; Stramska and Dickey, 1993). In estuaries, on continental shelves, and in the open ocean, physical processes appear to form the foundation (albeit a dynamic foundation) on which biological processes occur (Mann and Lazier, 1996). The pertinent physical and chemical characteristics of oceans that affect the nature and dispersal of marine life forms include (1) spatial scale, (2) properties of seawater as a life-support medium, (3) water temperature gradients, (4) mechanisms of water movement, and (5) marine ecosystem boundaries. (For more extensive and detailed analyses of marine processes, see Mann and Lazier, 1996).

2.1 Spatial Scale

The scale of the oceans and the ecosystems they contain is far larger than that of the land. The oceans cover approximately 70% of the earth's surface, but even more impressive is the difference in volume between marine and terrestrial environments. Though bacteria have been found in deep core samples, the vast majority of terrestrial life occupies a thin layer from a few meters (m) below the soil surface to 125 m above it (Norse, 1993). In contrast, marine organisms inhabit an environment that extends from the splash zone a few meters above the ocean surface to a maximum depth of more than 11,700 m (38,000 feet, or roughly 6.5 miles) in the Mariana Trench. The inhabited oceanic realm averages nearly 4,000 m in depth (Odum, 1971). Thus, though the oceans cover not quite three-quarters of the earth's surface area, they represent over 90% of the total volume of the biosphere, with only the small remainder contributed by land and air (Powers, 1995).

Although freshwater and marine ecosystems support similar species, the scale of the oceans dwarfs that of even the largest freshwater environments. Ocean basins are typically 10,000 kilometers (kms) wide and contain the largest biological communities (Mann and Lazier, 1996). Oceans contain 97% of all water in the hydrosphere; in comparison, freshwater lakes, rivers, and streams contain less than 0.2% (Goldman and Horne, 1983; Pierzynski et. al, 1994). With the exception of a few very large lakes (e.g., the Laurentian Great Lakes and Lake Baikal) and certain river basins (e.g., the Amazon River basin), freshwater systems are minute by comparison and are mainly small, shallow, and relatively isolated. In essence, freshwater bodies consist of small islands of water (individual ponds, lakes, streams, and rivers) separated by a sea of land and salt water (Hildrew et al., 1994). In turn, individual freshwater systems can be studied only at small spatial scales. Increasing spatial perspective rapidly leads one out into the surrounding landscape and into the groundwater. As a result, freshwater ecology at larger spatial scales can be viewed as merging with terrestrial ecology to form landscape ecology. In marine ecology, however, the spatial scale of one's approach can move from very small to very large while remaining fully oceanic (Hildrew et al., 1994).

2.2   Water as a Medium

Both the terrestrial and aquatic environments can be characterized as consisting of a solid base overlaid by fluid (air over ground-surface in one case, and water over sea-bottom in the other). As a life-support medium, however, water is qualitatively different from the atmosphere, and thus supports organisms with life history strategies quite distinct from those observed on land. Characteristics of water that affect the biological composition of marine communities include (1) density, viscosity, and hydrostatic pressure; (2) photic effects; and (3) chemistry.

2.2.1   Density, viscosity, and hydrostatic pressure

Water is 850 times more dense and 60 times more viscous than air (Strathmann, 1990). The much greater density of water compared with air provides mechanical support for a vast array of organisms, many of which live their lives suspended in this fluid medium free of contact with the ocean floor. The greater buoyancy and viscosity of water lowers the rate at which particles sink to the substrate and allows for the existence of a vital aquatic dispersal mechanism, planktonic dispersal (dispersal while floating in the water column) and a unique aquatic developmental strategy, planktotrophic development (feeding and larval growth while in the water column) that have no analogs in terrestrial ecosystems (Barnes and Hughes, 1988; Norse, 1993; Pasciak and Gavis, 1974). (See Chapter 3 for more details.)

Biological and ecological effects that result from physical and chemical influences vary depending on organism body size and the scale on which processes take place. For example, ocean water acts as a viscous liquid for organisms less than one millimeter in length, but loses this property for animals with greater size (Purcell, 1977; Strickler, 1984). Because of this high viscosity, the smallest marine organisms must depend on molecular diffusion for the transfer of nutrients and waste products. However, for larger animals, nutrients and wastes are moved more rapidly by turbulent diffusion and are unaffected by viscosity (Mann and Lazier, 1996). Thus, when considering the risk of introduced marine GEOs, it is essential to understand the scale on which adverse effects may occur and the physical and biological processes interacting at this scale.

Hydrostatic pressure also greatly influences distribution and abundance of marine life. At depths below 500 m, the partial pressures of gases dissolved in seawater are reduced, thereby affecting the biochemical reactions of organisms. In particular, dissolved oxygen concentrations decrease with depth so that when organisms adapted for life at or near the ocean surface are transported to deep water, they may suffer elevated mortality rates or revert to a dormant state. For example, natural blooms of phytoplankton tend to sink in the water column as they approach senescence (Mann and Lazier, 1996; Smayda, 1970). However, some organisms can remain as active in the depths as at the surface—viruses in particular seem unaffected by hydrostatic pressure.

2.2.2 Photic effects

Water absorbs light. Long wavelengths are absorbed in the first few meters; shorter wavelengths penetrate somewhat further (Jerlov, 1976). Beyond a certain depth, ranging from 100 to 1000 m depending on local meteorological and water clarity conditions, no light penetrates (Burchett, 1996). Therefore, the ocean, and aquatic systems in general, can be seen as divided into photic and aphotic zones.

As a result of the attenuation of photic energy as light passes through water, photosynthesis capable of exceeding the minimum energy required for respiration can take place only in the top 50 to 100 m of the surface waters of the ocean. Although the average depth of the ocean is 4000 m, the depth of the photic layer (approximately 100 m) is critical to open ocean biological processes (Mann and Lazier, 1996). Organisms that depend directly on photosynthesis, and those that in turn consume photosynthesizers, clearly have limitations as to how deep they are able to live in the water column. Surveys of marine ecosystems demonstrate that as light energy decreases so do the size, number, and population densities of pelagic communities (Angel, 1994). In general, ecosystem arrays in sub-photic zones (no light and/or dim light zones) are less dense and less intricately layered than those near the surface.

Nevertheless, many organisms have evolved unique life history strategies allowing them to inhabit the deeper ocean layers where light does not penetrate. Some marine creatures adapted to the lack of light move from the depths to the surface each night to take advantage of nutrients available in the photic zone, while others reside permanently in the aphotic regions. In the 1990s, growing recognition of a large abundance of deep sea fauna has emerged. For example, Grassle and Maciolek (1992) sampled 1,597 species of marine macroorganisms from the benthos collected off the east coast of North America at depths of 255 to 3,494 m and estimated that the global deep sea fauna may include as many as 10 million species—mostly polychaete worms, crustaceans, and mollusks (Reaka-Kudla, 1997).

2.2.3 Chemistry

The ocean medium has a composition of approximately 3.5% dissolved biogeochemical materials and 96.5% pure water (Barnes and Hughes, 1988). The concentration of dissolved nutrient chemicals affects species diversity and the abundance of organisms that live in particular habitats. Elements vital for support of life in the oceans include oxygen, carbon, phosphorus, nitrogen, silicon, and calcium. Generally these elements are utilized by living organisms in the photic layer and later redeposited into the deeper strata of the water column as these organisms die and decay.

Most open ocean seawater within several hundred meters of the surface is rich in the dissolved oxygen necessary for the support of aerobic microorganisms. However, physical processes can result in mixing that depletes essential levels of dissolved oxygen, significantly altering local communities. For example, upwellings of low-oxygen, nutrient-poor water can occur, resulting in local mass mortality of dependent life forms (Barber et al., 1985; Taunton-Clark and Shannon, 1988; see also the section on upwellings, below).

Changes in salinity also affect marine animals and microorganisms. Boundary zones between fresh, brackish, and salt water can present fluid barriers to the movements of organisms that are acclimated to one environment or the other. The mechanisms for salt control and ionic adjustment are so inflexible for a majority of aquatic organisms that they often are classified as stenohaline—restricted to a small range of salinity variation, either in salt or freshwater (Royce, 1984). However, certain macroorganisms, including salmon, sea trout, and eels, have adapted to tolerate a wide spectrum of salinity (Kleckner and McCleave, 1985, 1988; Royce et al., 1968). These organisms have the ability to adjust their osmotic properties to retain or expel salt depending on ambient conditions (Burchett, 1996).

Dissolved chemical particulates are continually added to the oceans by various mechanisms that may significantly affect local populations of marine organisms. For example, river systems carry eroded sediments and nutrients to estuaries and deltas leaving heavy local deposits on continental shelves (Barnes and Hughes, 1988). In the North Sea, studies have shown that coastal waters have a productivity almost twice as high as the offshore areas, apparently due to a large flux of nutrients from the sediments carried by rivers that flow into a tidally mixed water column (Fransz and Gieskes, 1984; Mommaerts et al., 1984).

The physical mechanisms that serve as the means for deposition and/or removal of chemicals in the oceans are numerous. First, wind currents contribute to chemical deposition by spreading terrestrial particulates over wide areas of the ocean surface. Second, geothermal vents on the ocean floor emit dissolved particulate matter from the earth's crust into the deep ocean, often supporting an array of deepsea fauna (Grassle, 1985; Jones, 1985; Tunnicliffe, 1992; Primack, 1993). Third, gas exchange at the interface of the atmosphere and ocean injects elements from the air into the ocean surface layer over its entire geographic range. Finally, in the hydrologic cycle, rain and other forms of precipitation deposit chemical elements from terrestrial and atmospheric sources into the seawater (Ward and Elliot, 1995). Thus, many of the physical processes in oceanic systems alter local chemical composition and significantly affect the composition of the biological community.

2.3 Water Temperature

Oceanic temperature variations caused by differential exposure to solar heating are a major cause of biogeographic diversity and stratification of the water column (Burchett, 1996). Temperature gradients also play a fundamental role in creating ocean currents, which in turn are largely responsible for the extensive opportunities for planktonic migration/dispersal available to marine organisms (see below).

Strong vertical mixing followed by stratification of the water column through solar heating leads to an increase in phytoplankton production (Legrende, 1981). Vertical mixing brings nutrients from deep water to the surface, where the formation of boundaries by stratification confines phytoplankton to a well lighted zone in which daily photosynthesis exceeds daily respiration (Gran, 1931; Sverdup, 1953). The driving force for vertical mixing is wind at the surface, while the factor producing stratification is the temperature differential from solar heating (Mann and Lazier, 1996). In an otherwise homogeneous water column, temperature variations of less than 0.05° Celsius (C) can produce differences in phytoplankton abundance of several orders of magnitude (Ryther and Hulbert, 1960). Research has also shown that some marine fish can detect water temperature changes as small as 0.03° C and move to areas with optimal temperatures (Owen, 1975; Royce, 1984). Further, the cooling of warm, near-surface waters in the winter causes convective mixing, bringing settled nutrients from the ocean depths back to the surface for utilization by free-floating plankton (Angel, 1994). In turn, nutrient availability affects growth of microorganisms and alters the balance between primary producers and those higher on the food chain. Thus, temperature differentials can have both direct and indirect effects on organisms at a particular site.

In general, relatively high temperatures create conditions for increased propagation by microorganisms, which can result in blooms of cyanobacteria, dinoflagellates, and microalgae. For example, during the summer of 1994, the water temperature in the southern Baltic Sea reached a nearly unprecedented 24° C. When the water temperature was at its highest, a toxic red tide occurred at this site for the first time in recorded history. Higher temperatures also tend to promote the transformation of marine bacteria from the "viable but nonculturable" stage to the culturable state (see Chapter 4).

Movements of tropical water masses that alter temperature can result in an associated shifting of marine boundaries, allowing many tropical species to temporarily occupy normally temperate waters. For example, in 1982 a warm water El Niño Southern Oscillation current (ENSO) shifted marine boundaries along the western coast of the United States, allowing species rarely observed so far north—including barracuda, marlin, ocean sunfish, sailfish, sea turtles, and certain sharks—to appear along the coast of Washington State. Ambient water temperatures rose from a normal range of 13° C to 15° C to as high as 18° C (Klee, 1991). During this event, the marine environment displayed drastic ecological changes as the El Niño inhibited normal upwelling of cooler, nutrient-rich water. The kelp forest canopy was reduced, and populations of herring, juvenile rockfish, anchovy, and salmon declined (Klee, 1991). Periodic ENSOs driven by temperature fluctuations are known to carry biotic propagules across the full width of the Pacific Ocean, causing dramatic, widespread population and community changes (Glynn, 1988; NRC, 1995).

Marine organisms display wide variation in temperature tolerance resulting in distinctive ecological ranges. Fish endemic to arctic and antarctic regions can survive in water temperatures as low as -2° C as a result of "antifreeze" proteins in their blood streams. In contrast, tropical surface waters, which support a rich diversity of marine life, commonly stabilize at temperatures around 30° C. Those fish with habitats outside the tropics and polar regions survive in a broad range of temperatures within these extremes. However, most free-swimming aquatic animals demonstrate a temperature preference. Many species migrate seasonally according to the warming and cooling of water (Royce, 1984). Therefore, temperature gradients may significantly affect the migratory patterns of marine organisms and/or the biogeographic regions in which the organisms are found (Burchett, 1996).

2.4 The Mechanisms of Water Movement

Patterns of oceanic movement are driven by astronomical effects, including the earth's rotation and the gravitational pull of the moon and sun. Because these influences affect the planet on a global scale, the ecological implications are profound. Organisms may be carried thousands of kilometers by transoceanic currents. In addition, forces that influence the earth on a planetary scale also spawn layered regional and local effects that interact in complex, dynamic ways.

2.4.1 The Coriolis effect

The earth's rotation strongly influences the flow of water across the three dimensions of ocean space (Semtner, 1995). The Coriolis effect, the inertial force that deflects any moving object from its path as the earth rotates beneath it (as viewed by an observer within the rotating system), is responsible for large scale patterns of atmospheric and oceanic movements (Mann and Lazier, 1996).

In the northern hemisphere, fluid flows are redirected to the right. This accounts both for the northeasterly flow of warm water currents originating in the tropics and for latitudinally distinct bands of global wind patterns, such as prevailing westerly winds and tropical trade winds. In the southern hemisphere the deviation of flow is reversed; wind and water currents are deflected to the left. Such flow patterns predominately determine the distribution of heat, salt, and nutrients—and marine organisms—in estuaries and on continental shelves (Mann and Lazier, 1996).

2.4.2 Currents

Currents are unidirectional flows of water, moving at speeds ranging from a few centimeters per second up to 5 m per second. Currents vary in size from the Gulf Stream, which is over 75 kilometers wide and flows thousands of miles along the eastern coast of North America as part of a huge clockwise North Atlantic rotation (see Plate 1, p. 53), to local currents that are only a few meters to several hundred meters wide. The paths followed by many currents are relatively regular and predictable (Mann and Lazier, 1996). These flows can mix water of distant parts of one ocean, or even water from different oceans (de Bilj, 1987). Consequently, currents carry organisms great distances and, therefore, constitute a mechanism for global dispersal of marine organisms across oceanic basins (Coelho, 1985; Rowell and Trites, 1985; Scheltema, 1966, 1971a, 1971b, 1986).

The forces primarily responsible for producing the major ocean currents are temperature, pressure variations, and rotational effects. For example, persistent low pressure systems over the Atlantic Ocean to the north and south of the equator combine with the easterly winds near the equator and westerly winds in the middle latitudes to stimulate a great clockwise circulation in the North Atlantic and a counterclockwise rotation in the South Atlantic (de Bilj, 1987). Wind-driven movements of ocean surface water result from momentum transference at the interface of the two fluids. Slower circulatory flows powered by temperature and water density variations, or thermohaline effects, transfer energy to the ocean depths and result in currents that move the entire water column (Burchett, 1996). Organisms at all depths of the ocean from the surface to the sea floor are affected by these large scale circulations.

2.4.3 Eddies, rings, and gyres

The western boundary currents in both the northern and southern hemispheres tend to be fast, deep, and relatively narrow, concentrating a large energy source in a small cross-section of water (e.g., Gulf Stream and Kuroshio Current). These meandering currents, which are subject to many types of largely unpredictable forces, give rise to eddies, rings, and gyres of various kinds (Mann and Lazier, 1996; Swallow, 1976).

The term "eddy" usually connotes lower amplitude variations that are found throughout the ocean. Ocean eddies are complex, countercurrent whirlpools that typically rotate once in a 10 to 30-day time period and can range in size from tens to hundreds of kilometers in width (Robinson, 1983).

"Rings" can be created by eddies that break off and form distinct ring-like, independent entities separate from the originating current. For example, rings are formed from meanders of the Gulf Stream and persist as distinct entities for many months. Cold-core rings form on the south side of the Gulf Stream, and warm-core rings form on the north (see Plate 2, p.53). Once separated from the Gulf Stream, these rings move in a southwesterly direction, eventually reuniting with the main current or losing their identity by diffusing into the surrounding water (Mann and Lazier, 1996). Cold-core rings have been shown to support highly productive communities. For example, primary production in cold-core rings from the Sargasso Sea was estimated to be about 50% higher than in the sea itself, yet the rings occupied only 10% of the Sargasso Sea at any one time (Ring Group, 1981). Empirical studies have also shown that warm-core rings, although having the opposite internal structure of cold-core rings, experience bursts of relatively high biological production associated with the vertical transport of nutrients (Hitchcock et al., 1987; Tranter et al., 1980).

"Gyres" are circular currents shaped by features of the ocean floor. Gyres are found on a wide range of spatial scales. At the scale of an entire ocean basin, for example, the North Atlantic Subtropical Gyre is made up of the Gulf Stream, the North Atlantic Current, the Canary Current, and the

North Equatorial Current. An example of a gyre at a smaller scale is the Georges Bank Gyre, the continuous current flow over the Georges Bank in the Gulf of Maine (Mann and Lazier, 1996).

Eddies, rings, and gyres are the principal mechanisms for mixing in the oceans (Angel, 1994). Therefore, they can profoundly affect the distribution of pelagic organisms by influencing ambient water temperature and local availability of nutrient supply, or by direct transportation of entire communities from one site to another. In particular, gyres are key mechanisms for carrying plankton and microorganisms over long distances and for mixing otherwise discrete populations. The European eel, *Anguilla*, and the American eel, *Anguilla rostrata*, utilize the North Atlantic Gyre for breeding and for transportation into rivers to feed and mature. Successful larvae ride the Western Boundary and North Atlantic currents while breeding adults follow separate return flows (Kleckner and McCleave, 1985, 1988). Several species of Pacific salmon swim downstream with the Alaskan Gyre, which apparently aids their feeding and migration efforts in the open sea (Royce et al., 1968).

2.4.4 Tides

Tidal effects result primarily from an interaction of three forces: the gravitational attraction of the moon, the gravitational attraction of the sun, and the centrifugal force that results from the revolution of the earth around the common centers of gravity of the earth and these two bodies (Mann and Lazier, 1996). The strength of tidal forces varies seasonally depending on interference patterns between solar tides and lunar tides. Lunar effects are more than twice as strong as solar effects because of the moon's astronomical proximity to the earth. Local tidal variations stem from topographical features of continental coastline formations, latitudinal Coriolis influences, and other factors (Hendershott, 1981). Tidal lifts range in size from several centimeters in the open ocean to over 15 m in the Bay of Fundy on Canada's eastern coast.

Tides are one of the fundamental forces affecting the distribution of marine life. As they inundate intertidal zones and then retreat to let the exposed areas partially dry, these most familiar twice-daily ebbs and flows are a mechanism for mixing terrestrial and marine biota. Tides can generate currents in the water that mix lower layers, affecting the distribution and community structure of organisms. Littoral microorganisms that otherwise would settle to the sea bottom can become stirred up by tidal forces and intermixed with pelagic bacteria and microalgae. Patterns of zonation by intertidal organisms have been documented in numerous studies (Connell, 1961; Lubchenco, 1978; Paine, 1966). Further, the breeding grounds of certain marine fish such as herring have been documented to exist in distinct areas of strong tidal mixing (Iles and Sinclair, 1982).

2.4.5 Upwellings and downwellings

Physical factors determining habitat fertility are quite different in terrestrial and marine environments. Nutrients essential for plant growth in terrestrial systems are generated locally from the detritus of previous generations. In the oceans, decaying matter tends to sink and leave the sunlit photic zone where phytoplankton grows. Nutrients essential for phytoplankton growth are thus unavailable unless some physical mechanisms bring the nutrients back to the surface (Mann and Lazier, 1996). Upwellings serve as one such physical mechanism by which nutrients are delivered into the photic zone to support marine life forms.

Upwellings and downwellings are vertical currents that result from an interplay of winds, surface motion, and temperature gradients within the water column. Warm oxygenated surface water may be carried to the depths or, conversely, cold nutrient-rich waters may be brought to the surface (Chapman and Shannon, 1985; Shannon, 1985). Depending on prevalent meteorological conditions, these movements can be continuous, seasonal, or intermittent (Blanton et al., 1984; Rasmusson and Wallace, 1983; Wooster et al., 1976). Upwellings and downwellings generally transport water at lower speeds than surface currents, moving at speeds of only a few centimeters per day.

Vertical movements of ocean water can have important ecological effects (Cushing, 1969; Cushing, 1971; Denny, 1988). Locally, an upwelling can stimulate the growth of primary producers such as phytoplankton. A phytoplankton bloom, in turn, may cause explosive growth in secondary feeders, such as zooplankton, and result in a cascade of intertrophic effects that move upward through the food web (Chavez and Barber, 1987; Trumble et al., 1981; Walsh et al., 1980).

2.4.6 Waves and aerosols

Waves are created by meteorological effects, tidal action, and, occasionally, even by tectonic disturbances. Waves and the energy contained in them can traverse thousands of miles across oceans. Wave size depends upon wind speed, duration, and the distance over which wind flow is sustained (Burchett, 1996). Waves impinging on coastlines can significantly alter shallow marine habitats. For example, when the tidal current generated by an ebbing tide flows over the edge of the continental shelf, it forms waves on the downstream side (Halpern, 1971; Chereskin, 1983). Nearshore habitats can be created and destroyed as waves deposit materials, oxygenate the water, and deliver essential nutrients (Burchett, 1996). Further, waves agitate surface waters and serve as a transport mechanism for various marine organisms (Shanks, 1983, 1985; Shanks and Wright, 1987; Zeldis and Jillett, 1982). Harsh conditions from mechanical wave action require that all plants and animals living in areas of wave influence be well adapted for survival in such dynamic conditions (Burchett, 1996).

Aerosols, which often contain microorganisms, are created by breaking waves and bubbling action in the water column. Bubbles that rise to the surface and then burst can inject microorganism-laden droplets as high as 20 cm into the atmosphere (Lighthart and Stetzenbach, 1994). Breaking waves, both at the shoreline and in the open ocean, can throw foam several meters into the air. The fine wind-borne aerosols thus created can remain aloft for extended periods and may cover great distances before returning to mix with surface waters at a new location. Aerosols that carry microorganisms (bioaerosols) can significantly alter biological communities and have been implicated as a mechanism for spreading pathogens lethal to coral throughout the southern seas. Similarly, bivalve and crustacean larvae are also known to travel via aerosols (ABRAC, 1995).

2.5 Marine Ecosystem Boundaries

In contrast to most terrestrial and freshwater habitat frontiers, marine boundaries are dynamic, often moving on a time scale of days or even hours as ocean waters circulate (Mann and Lazier, 1996; Norse, 1993). Satellite photography has recorded swirls of phytoplankton and the other organisms they support shifting many kilometers on a daily basis (Norse, 1993). In addition to the influence of major currents, minor fluctuations in factors such as temperature, salinity, dissolved oxygen, light gradients, and surface winds all enhance the dynamism of marine ecosystem boundaries.

Though boundaries exist between oceanic habitats, they are fluid and permeable, unlike the more severe and exclusive demarcations in terrestrial systems. Short-term biotic shifts following dynamic boundary adjustments are far more common in the marine environment than on land and often undermine attempts of oceanographers to establish fixed biogeographic ranges for marine organisms (Mann and Lazier, 1996; Norse, 1993). Unlike landmass and freshwater environments, the oceans are fully interconnected. Consequently, marine species, on average, have larger ranges than land organisms (Norse, 1993).

Responding to the dynamic nature of ocean water boundaries, natural selection may promote long distance dispersal to minimize negative effects, such as high mortality, that often result from local environmental perturbations. Dynamic ecosystem boundaries may act as a factor supporting widespread use among marine organisms of the planktonic dispersal strategy. For instance, although the deep sea may offer no suitable habitat for certain tropical invertebrates, species can traverse these areas to successfully establish populations on both sides of an ocean (Scheltema, 1966, 1971a, 1971b, 1986).

3. BIOTIC FEATURES

Biological characteristics of the ocean environment also affect the composition, structure, abundance, and distribution of marine organisms and systems. In this section, we consider marine life history strategies, including reproductive modes and dispersal strategies, and primary production.

3.1 Marine Life History Strategies

The life history of an organism can be defined as its lifetime pattern of growth, differentiation, storage, and reproduction (Begon et al., 1990). Life history strategies reflect the genetic makeup of an organism, its environment, and the interaction between these two factors. The life history of an organism can be viewed as an expression of a strategy for growth, reproduction, survival, and dispersal. Organisms may spend varying amounts of time in each phase of growth and differentiation prior to reproducing. They may reproduce once or repeatedly, yielding varying quantities of different sizes of eggs/larvae, and utilize numerous methods for introducing their reproductive products into the environment.

Characteristics of Marine Ecosystems Relevant to Uncontained 45
Applications of Genetically Engineered Organisms

3.1.1 Reproductive modes

Marine organisms exhibit a wide array of reproductive modes. One common method utilized by ocean species is dioecy, in which male and female reproductive organs are in separate individuals, thus requiring sperm transfer. Non-dioecious modes of reproduction are exhibited as well; although variable across species, these can be classified into two general categories: hermaphroditism and asexual reproduction (ABRAC, 1995).

Hermaphroditic organisms possess both male and female organs and may reproduce through cross- or self-fertilization. Hermaphroditic organisms may display both female and male characteristics at one time, termed simultaneous hermaphroditism, or may experience sequential hermaphroditism (Dawley, 1989; Sadovy and Shapiro, 1987; Sunobe and Nakazono, 1993). Various fish species (including the sea breams, *Rivulus marmoratus* and *Sparidae* spp.) and some plants exhibit simultaneous hermaphroditism (Soto et al., 1992; Turner et al., 1992; Buxton and Garret, 1990). Examples of sequential hermaphrodites include the northern shrimp (*Pandalus*) and reef-dwelling fish such as the wrasses (*Labridae*), parrotfish (*Scaridae*), and gobies (*Gobiidae*) (Royce, 1984; Warner, 1978, 1984, 1988; Wootton, 1992). Of the 5,600 known mollusk genera, 40% are estimated to be either simultaneous or sequential hermaphrodites (Heller, 1993).

In contrast, asexual reproduction is defined as the process by which an organism reproduces without the formation of a zygote (Levington, 1995). Types of asexual reproduction vary according to the biology of the individual species. For example, marine phytoplankton (e.g., diatoms) and colonial animals (e.g., sponges, coelenterates, and bryozoans) reproduce through fission. Seaweeds and corals utilize fragmentation. Many other marine organisms, including members of the fish families *Poeciliidae*, *Atherinidae*, *Cyprinidae*, and *Cobitidae*, display parthenogenesis (development from an unfertilized egg, seed, or spore) (Dawley, 1989; Moore, 1984; Levington, 1995).

3.1.1.1 High fecundity.
Fecundity can de defined as the number of ripe eggs/seeds produced by an organism during a given reproductive episode (Silvertown, 1982; Wootton, 1992). The majority of marine organisms are highly fecund, producing larger numbers of eggs than most terrestrial species (Leiby, 1984; Norse, 1993). Examples of highly fecund marine fish include moonfish (*Mola*

mola: 3 billion eggs), tarpon (*Megalops atlanticus*: 12 million), Atlantic cod (*Gadus callarius*: 9 million), haddock (*Melanogrammus aeglefinus*: 3 million), Pacific halibut (*Hippoglossus stenolephis*: 2.7 million), winter flounder (*Pseudopleuronectes americanus*: 1.5 million), hake (*Merluccius*: 1 million), and mackerel (*Scomber scrombus*: 0.5 million) (Herald, 1961; Hildebrand, 1963; Marshall, 1966; Owen, 1975). In comparison, terrestrial egg-laying vertebrates usually produce fewer than 100 eggs per episode.

Sessile marine invertebrates, such as oysters, also display extremely high levels of fecundity, expelling tens of millions of eggs in a single reproductive event (Royce, 1984). For example, *Crassostrea virginica*, an estuarine oyster, can produce up to 70 million eggs in a single spawning (Norse, 1993).

Fecundities exceeding 10 million are common among free-living aquatic animals, but rare for terrestrial ones (Strathmann, 1990). Comparisons between marine and freshwater species also confirm that, on average, ocean dwelling fish are more fecund than freshwater ones (Owen, 1975).

3.1.1.2 Small egg/larva size.

Reproductive strategies often involve trade-offs due to limited energy supply (Begon et al., 1990). One trade-off is between fecundity and egg

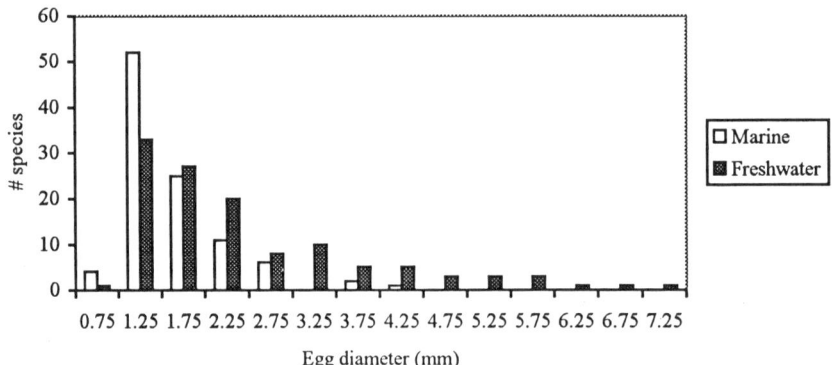

Figure 1. Egg size for marine and freshwater species of stream-spawning salmonids in northern temperate waters (Wootton, 1979; Elgar, 1990). Marine species tend to produce smaller eggs than freshwater species (from Wootton, 1992).

size. Organisms with high fecundity tend to produce smaller eggs (Salthe, 1969; Begon et al., 1990; Elgar, 1990). For example, sexually mature cod and salmon with similar body sizes have different fecundity levels and corresponding differences in egg size. Salmon produce a few thousand relatively large eggs (4 to 7 millimeters in diameter) while cod spawn several hundred thousand smaller eggs (1.5 mm in diameter) (Wootton, 1992). Marine fish, which on average are more fecund than freshwater fish, tend to produce smaller eggs (see Figure 1, above). For example, among stream-spawning salmonids, marine species tend to produce smaller eggs than freshwater species (Wootton, 1979; Elgar, 1990).

3.1.2   Dispersal strategies

Striking differences occur between dispersal mechanisms utilized in the oceans and those used in terrestrial/freshwater systems. Marine organisms often display unique life history strategies to utilize ocean dynamics for dispersal to suitable habitats. These strategies correlate with the characteristics of local marine ecosystems that support adult populations. For instance, spawning reef fish tend to form aggregations near locations where there is substantial current flow, or at sites that are down-current and seaward of the reef. Yearly spawning is often synchronized with current flow or tidal changes in these systems. Species such as the Atlantic parrotfish (*Sparisoma rubripinne*), the spotted goatfish (*Pseudopeneus maculatus*), and the striped parrotfish (*Scarus croicensis*) deposit eggs into the water column at the peak of the current, which consequently flushes the eggs away from predators on the reef (Randall and Randall, 1963; Colin and Clavijo, 1978; Johannes, 1978; Barlow, 1981; Colin, 1978). Other marine species migrate seaward at the time of egg-laying. They then rely on planktonic dispersal to carry eggs inshore to the estuaries and coastal areas that serve as nursery grounds. For example, *Spratus spratus*, a valuable commercial fish in Norway and Sweden, and cod (*Gadus morhua*) migrate out to sea and rely on currents to distribute eggs into suitable nursery areas close to shore (Lindquist, 1968; Ellerston et al., 1981).

Some fish species deposit eggs in sand along coastal waters, utilizing the subsequent high tide to flush the eggs out to sea, while others swim ashore at high tide when spawning. When the water recedes, taking the fish back into the ocean, the dispersed eggs become immersed in damp vegetation or sand. Other survival mechanisms may then come into play.

For example, fish of the order *Atheriniformes* produce eggs that can withstand desiccation and hatching delays for prolonged periods (Harrington, 1959; Koenig and Livingston, 1976; Reynolds and Thompson, 1974; Taylor et al., 1977). When the eggs are once again immersed by a subsequent high tide, hatching takes place and larvae are carried into ocean waters by waves (Harrington, 1959; Taylor et al., 1977). These fish species may take advantage of planktonic dispersal to reduce the risk of predation on eggs in the estuary (see below).

3.1.2.1 Planktonic dispersal and planktotrophic development.

It is common for marine plants and animals to inject their reproductive products into the water column where they become temporary plankton (free-floating propagules) called meroplankton (Strathmann, 1990). The majority of marine organisms, including free-swimmers, bottom-dwellers, and those found within the water column, release eggs, larvae, or spores for dispersal in this manner (Leiby, 1984; Norse, 1993; Strathmann, 1990; Thorsen, 1950). Figure 2 shows the large number of teleost reef fish families (i.e., fish with bony skeletons) that release eggs as meroplankton.

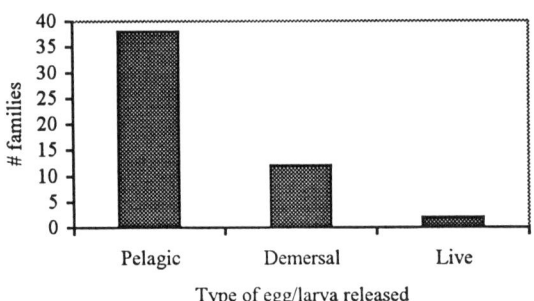

Figure 2. Distribution of the types of eggs/larvae released among reef teleost fish, expressed as the number of families that include species displaying release of a particular type of egg/larva (pelagic = release into the water column; demersal = deposition on the bottom; live = live birth). Families may be represented in more than one category (after Thresher, 1984).

Once present in the water column, planktonic larvae are capable of long distance dispersal utilizing transoceanic currents (Scheltema, 1966, 1971a, 1971b, 1986). The time spent in the planktonic stage usually ranges from hours to weeks, but can extend to more than a year (Norse, 1993). For example, some Pacific echinoderm larvae spend as many as 36 weeks drifting as plankton (Barnes and Hughes, 1988). Similarly, larvae of the Caribbean spiny lobster spend nine months or more as plankton, and may traverse the entire width of the open ocean pelagic ecosystems of the North Atlantic (Farmer et al., 1989).

In addition, the characteristics of the ocean allow for the existence of holoplankton, a unique category of organisms for which there is no counterpart on land. These organisms, which include bacterioplankton, zooplankton, and phytoplankton, live permanently near the surface and may serve as vital nourishment for planktotrophic developers during their extended larva stage (Barnes, H., 1956; Barnes, R., 1991; Chipperfield, 1953; Levington, 1982). Thus, marine larvae utilizing planktotrophic development (i.e., feeding and larval growth while dispersing in the water column), are not limited to energy provided solely by their parents; they can take advantage of holoplankton as an alternative energy source, and continue to develop while dispersing, a strategy unavailable to passively dispersing terrestrial propagules (Crisp, 1976; Begon et al., 1990). Figures 3 and 4 (below) present the occurrence of planktotrophy among 30 species of nudibranch mollusks and three families of deepsea snails. In these representative groups, planktotrophy is by far the dominant form of larval development.

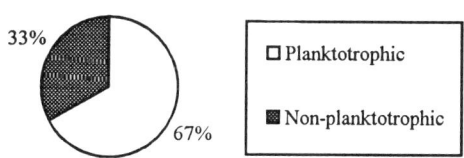

Figure 3. Percentage of nudibranch mollusk species utilizing planktotrophic larval development from a representative sample of 30 species, as presented by Todd and Doyle (1981).

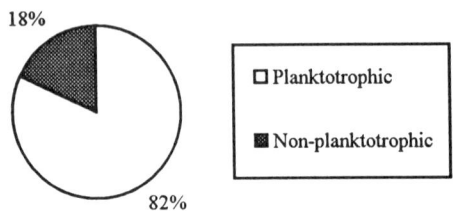

Figure 4. Frequency of planktotrophy in the gastropod families *Neritimorpha*, *Caenogastropoda*, and *Heterostropha*. Once believed to be rarely utilized among deep sea snails, planktotrophy has recently been recognized as a common form of larval development (from Bouchet and Waren, 1995).

Although long range dispersal of reproductive products is not limited to the marine environment, it is far more common in the oceans than on land or in freshwater systems (Strathmann, 1990). Terrestrial organisms such as fungal spores, windblown seeds, pollen, and ballooning spiderlings are, in essence, "aeroplankton" utilizing atmospheric currents to carry them to new locations (Begon et al., 1990; Norse, 1993). However, gravity and a limited energy supply reduce the effective dispersal range of land-based organisms. For example, many wind-dispersed seeds fall close to the parent; there is a marked decrease in density as distance from the source plant increases (Begon et al., 1990; Marchand and Roach, 1980; Sheldon and Burrows, 1973; Silvertown, 1982).

Planktonic dispersal of eggs is also uncommon among freshwater species in ponds and lakes. Land barriers and the lack of strong currents reduce opportunities for long range transportation of propagules. Habitat quality varies much less within lakes than in the ocean, resulting in little selective advantage for long distance dispersal in freshwater (Hughes, 1980). Finally, unidirectional riverine currents are disadvantageous for successful dispersal since they often flush eggs downstream to unfavorable habitats.

3.1.2.2 Site selection.

The planktotrophic larvae of numerous marine species are influenced by environmental cues that induce active settlement (Barnes and Hughes, 1988; Chia and Rice, 1978). For example, chemoreceptive response to the presence of adults has been documented in the case of barnacles and oysters (Hughes, 1980) and abalone (Morse and Morse, 1988; Morse, 1991). At sea, an organism may be able to postpone settlement until certain environmental cues are encountered that indicate a hospitable location (Chia and Rice, 1978). Thus, marine larvae often can avoid poor habitat, while seeds (although possessing the ability to abstain from germination until good growing conditions exist) lack the ability to influence habitat selection (Strathmann, 1990). Terrestrial "aeroplankton" are passive dispersers with no power to determine where they will fall (Begon et al., 1990).

3.2 Primary Production

One major difference between the terrestrial and marine environments concerns primary producers (Ray, 1976). On land, primary producers (mainly vascular plants such as grasses and trees) constitute a majority of the biomass and are often relatively large, long-living organisms (Steele, 1985, 1991; Steele et al., 1989). Except in cases where macrophytes (e.g., seaweeds and seagrasses) occur along the shoreline in littoral zones, primary producers in lakes and oceans are quite small and are outweighed by their consumers on a per individual basis (Norse, 1993). Such primary producers are often microscopic, short-lived phytoplankton that reproduce rapidly and thus display high turnover rates.

With high turnover rates, phytoplankton populations respond quickly to environmental changes and fluctuations in populations (Mann and Lazier, 1996). In contrast, long-lived primary producers on land (e.g., trees), respond more slowly to environmental fluctuations (Norse, 1993; Steele, 1985, 1991; Steele et al., 1989). High turnover rates for marine primary producers can increase the vulnerability of consumer populations to severe disturbances of even relatively short duration (Norse, 1993). For example, a pulse of nutrient enrichment (e.g., from an upwelling) can trigger a phytoplankton bloom, thereby affecting the growth and reproduction of zooplankton consumers and organisms in the higher trophic levels connected in the food web.

4. CONCLUSION

In 1995, the Committee on Biological Diversity in Marine Systems of the National Research Council concluded that "marine and terrestrial ecosystems differ in significant ways that suggest the ocean may respond to human perturbations in a fundamentally different manner from the land" (NRC, 1995). The overview of key physical and biological processes discussed above supports this contention. Marine systems support complex food webs on spatial scales that are much larger than seen on land or in all but the largest freshwater lakes and rivers. Currents, eddies, rings, gyres, and tides all significantly affect marine biotic communities by altering local nutrient levels, influencing temperature regimes, shifting ecosystem boundaries, and providing the means for long distance migration and dispersal. Life history strategies, including high fecundity levels, small egg/larvae size, planktotrophic development, and a variety of planktonic dispersal mechanisms, play a greater role in marine systems. Marine life histories, adapted to the dynamism and continuity of ocean environments, provide greater opportunities for rapid reproduction and wide dispersal of biotic propagules than are possible in terrestrial/freshwater ecosystems. The implications of these differences for assessing risks associated with marine transgenic organisms must be considered. These issues are discussed further in Chapters 3 and 4.

Characteristics of Marine Ecosystems Relevant to Uncontained Applications of Genetically Engineered Organisms

Plate 1. Satellite imagery displaying water temperature differentials associated with Gulf Stream flows in the western North Atlantic (http://daac.gsfc.nasa.gov).

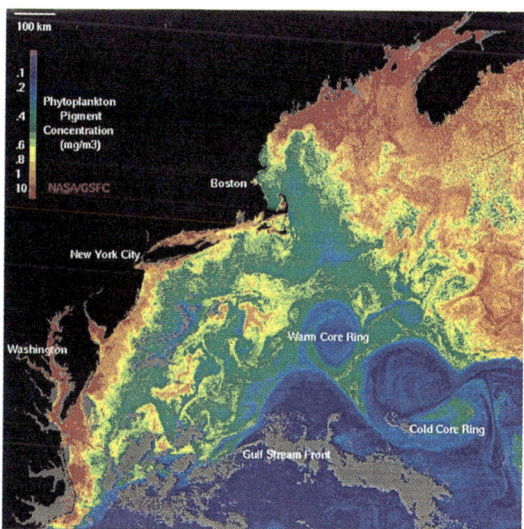

Plate 2. Satellite imagery displaying phytoplankton concentrations along the northeastern coast of North America. Among the factors affecting phytoplankton distribution are terrestrial runoff and Gulf Stream water temperature gradients (http://daac.gsfc.nasa.gov).

REFERENCES

Agricultural Biotechnology Research Advisory Committee (ABRAC). 1995. Performance Standards for Safely Conducting Research with Genetically Modified Fish and Shellfish. Document No. 95-04. Washington, DC: United States Department of Agriculture.

Angel, M.V. 1994. Long-term, large-scale patterns in marine pelagic systems. In: P.S. Giller, A.G. Hildrew and D.G. Raffaelli (eds.). Aquatic Ecology: Scale, Pattern and Process. London: Blackwell Scientific Publications, pp. 403-439.

Barber, R.T., F.P. Chavez and J.E. Kogepschatz. 1985. Biological effects of El Niño. In: M. Vegas (ed.). Seminario Regional Ciencas Tecnologia y Agression Ambiental: El Fenomeno El Niño. Lima, Peru: Contec Press.

Barlow, G.W. 1981. Patterns of parental investment, dispersal and size among coral-reef fishes. Environmental Biology of Fishes 6:65.

Barnes, R.S.K. 1991. Reproduction, life histories and dispersal. In: R.S.K. Barnes and K.H. Mann (eds.). Fundamentals of Aquatic Ecology. London: Blackwell Scientific Publications.

Barnes, H. 1956. *Balanus balanoides* (L.) in the Firth of Clyde: the development and annual variation of the larval population and the causative factors. Journal of Animal Ecology 25:72-84.

Barnes, R.S.K. and R.N. Hughes. 1988. An Introduction to Marine Ecology. 2nd Edition. Boston: Blackwell Scientific Publications.

Begon, M., J.L. Harper and C.R. Townsend. 1990. 2nd Edition. Ecology: Individuals, Populations and Communities. Boston: Blackwell Scientific Publications.

Blanton, J.O., L.P. Atkinson, F.F. de Castillejo and A.L. Montero. 1984. Coastal upwelling off the Rias Bajas, Galicia, Northwest Spain. 1. Hydrographic studies. Rapp. P.-v. Reun. Cons. Int. Explor. Mer. 183:79-90.

Bouchet, P. and A. Waren. 1994. Deep-sea gastropod larvae. In: C.M. Young and K.J. Eckelbarger (eds.). Reproduction, Larval Biology and Recruitment of the Deep-Sea Benthos. New York: Columbia University Press.

Burchett, M. 1996. Oceanography and marine biology. In: G. Waller (ed.). Sea Life: A Complete Guide to the Marine Environment. Washington, DC: Smithsonian Institution Press.

Buxton, C.D. and P.A. Garrett. 1990. Alternative reproductive styles in seabreams (*Pisces*: *Sparidae*). In: M.N. Burton (ed.). Alternative Life History Styles of Fishes. 28:113-124.

Chavez, F.P. and R.T. Barber. 1987. An estimate of new production in the equatorial Pacific. Deep-Sea Research 34:1229-1243.

Chapman, P. and L.V. Shannon. 1985. The Benguela ecosystem, Part 2. Chemistry and related processes. Oceanography and Marine Biology Annual Review 23:183-251.

Chereskin, T.K. 1983. Generation of internal waves in Massachusetts Bay. *Journal of Geophysical Research* 88:2649-2661.

Chia, F.S. and M.E. Rice. 1978. Settlement and Metamorphosis of Marine Invertebrate Larvae. New York: Elsevier Scientific Publishers.

Chipperfield, P.N. 1953. Observations of the breeding and settlement of *Mytilus edulis* (L.) in British waters. Journal of the Marine Biological Association of the United Kingdom 32:449-476.

Coelho, M.L. 1985. Review of the influence of oceanographic factors on cephalopod distribution and life history. NATO Sci. Coun. Studies 9:47-57.

Colin, P.L. 1978. Daily and summer-winter variation in mass spawning of the parrotfish, *Scarus croicensis*. Fish Bulletin 76:117.

Colin, P.L. and I.E. Clavijo. 1978. Mass spawning by the spotted goatfish, *Pseudopeneus maculatus*. Bulletin of Marine Science 28:780.

Connell, J.H. 1961. The influence of interspecific competition and other factors on the distribution of the barnacle, *Chthamalus stellatus*. Ecology 42:710-723.

Crisp, D.J. 1976. The role of the pelagic larva. In: P. Davies Spencer (ed.). Perspectives in Experimental Zoology. Oxford: Pergamon Press.

Cushing, D.H. 1969. Upwelling and fish production. F.A.O. Fish. Tech. Paper 84:38 pp.

Cushing, D.H. 1971. Upwelling and the production of fish. Advances in Marine Biology 9:255-334.

de Bilj, H.J. 1987. The Earth: A Physical and Human Geography. New York: John Wiley and Sons, Inc.

Dawley, R.M. 1989. An introduction to unisexual vertebrates. In: R.M. Dawley and J.P. Bogart (eds.). Evolution and Ecology of Unisexual Vertebrates. Bulletin 466. Albany, NY: New York State Museum.

Denny, M.W. 1988. Biology and the Mechanics of the Wave-Swept Environment. Princeton, NJ: Princeton University Press.

Elgar, M.A. 1990. Evolutionary compromise between a few large and many small eggs: comparative evidence in teleost fish. Oikos 59:283-287.

Ellerston, B., P. Solemdal, T. Stromme, S. Sunby, T. Tillseth, T. Westgard and V. Oiestad. 1981. Spawning period, transport and dispersal of eggs from the spawning area of Arcto-Norwegian cod (*Gadus morhua* L.). Cons. Inter. Explor. Mer. 178:260.

Farmer, M.W., J.A. Ward and B.E. Luckhurst. 1989. Development of spiny lobster (*Panulirus argus*) phyllosoma larvae in the plankton near Bermuda. In: G.T. Waugh and M.H. Goodwin (eds.). Proceedings of the 39th Annual Gulf and Caribbean Fisheries Institute. Charleston, South Carolina.

Fransz, H.G. and W.W.C. Gieskes. 1984. The unbalance of phytoplankton and copepods in the North Sea. Rapp. P. -v Reun. Cons. Int. Explor. Mer. 183:218-225.

Glynn, P.W. 1988. El-Niño-Southern Oscillation 1982-83: Nearshore population, community, and ecosystem responses. Annual Review of Ecological Systems 19:309-345.

Golman, C.R. and A.J. Horne. 1983. Limnology. New York: McGraw-Hill.

Gran, H.H. 1931. On the conditions for the production of plankton in the sea. Rapp. P.-v. Reun. Cons. Int. Explor. Mer. 75:37-46.

Grassle, J.F. 1985. Hydrothermal vent animals: distribution and biology. Science 229:713-717.

Grassle, J.F. and N.J. Maceilok. 1992. Deep-sea species richness: regional and local diversity estimates from quantitative bottom samples. American Naturalist 139:313-341.

Halpern, D. 1971. Observations on short-period internal waves in Massachusetts Bay. Journal of Marine Research 29:116-132.

Harrington, R.W. 1959. Delayed hatching in stranded eggs of marsh killifish, *Fundulus confluentus*. Ecology 40:430.
Hartline, B.K. 1980. Coastal upwelling: physical factors feed fish. Science 208:38-40.
Heller, J. 1993. Hermaphroditism in molluscs. Biological Journal of Linnean Society 48(1):19-42.
Hendershott, M.C. 1981. Long waves and ocean tides. In: B. A. Warren and C. Wunch (eds.). Evolution of Physical Oceanography. Cambridge, MA: M.I.T. Press.
Herald, E.S. 1961. Living Fishes of the World. Garden City, NY: Doubleday.
Hildebrand, S.F. 1963. Family *Elopidae*. In: Y.H. Olsen (ed.). Fishes of the Western North Atlantic. New Haven, CT: Yale University Press.
Hildrew, A.G., D.G. Raffaelli and P.S. Giller. 1994. Introduction. In: P.S. Giller, A.G. Hildrew and D.G. Raffaelli (eds.). Aquatic Ecology: Scale, Pattern and Process. London: Blackwell Scientific Publications, pp. ix-xiii.
Hitchcock, G.L., C. Langdon and T.J. Smayda. 1987. Short-term changes in the biology of a Gulf Stream warm-core ring; phytoplankton biomass and productivity. Limnol. Oceanography 32:919-928.
Hughes, R.N. 1980. Strategies for survival of aquatic organisms. In: R.S.K. Barnes and K.H. Mann (eds.). Fundamentals of Aquatic Ecosystems. Boston: Blackwell Scientific Publications.
Iles, T.D. and M. Sinclair. 1982. Atlantic herring: stock discreetness and abundance. Science 215:627-633.
Jerlov, N.J. 1976. Marine Optics. Elsevier Oceanography Series 14. Amsterdam: Elsevier.
Johannes, R.E. 1978. Reproductive strategies of coastal marine fishes in the tropics. Environmental Biology of Fishes 3:65.
Jones, M.L. (ed.). 1985. Hydrothermal vents of the Eastern Pacific: an overview. Bulletin of the Biological Society of Washington 6, 545 pp.
Kleckner, R.C. and J.D. McCleave. 1988. The northern limit of spawning by Atlantic eels (*Anguilla* spp.) in the Sargasso Sea in relation to thermal fronts and surface water masses. Journal of Marine Research 46:647-667.
Kleckner, R.C. and J.D. McCleave. 1985. Spatial and temporal distribution of American eel larvae in relation to the North Atlantic Ocean current systems. Dana 4:67-92.
Klee, G.A. 1991. Marine Resource Management. In: Conservation of Natural Resources. Englewood Cliffs, NJ: Prentice-Hall.
Koenig, C.C. and R.J. Livingston. 1976. The embryological development of the diamond killifish, *Adinia xenica*. Copeia 435.
Leiby, M.M. 1984. Life history and ecology of pelagic fish eggs and larvae. In: K.A. Steidinger and L. M. Walker (eds.). Marine Plankton Life Cycle Strategies. Boca Raton, FL: CRC Press.
Legrende, L. 1981. Hydrodynamic control of marine phytoplankton production: the paradox of stability. In: J.C.J. Nihoul (ed.). Ecohydrodynamics. Proceedings of the 12th International Liege Colloquium on Ocean Hydrodynamics. Amsterdam: Elsevier.
Levington, J.S. 1995. Marine Biology: Function, Biodiversity, Ecology. New York: Oxford University Press.
Levington, J.S. 1982. Marine Ecology. Englewood Cliffs, NJ: Prentice-Hall Inc.
Lewis, M.R., J.J. Cullen and T. Platt. 1983. Phytoplankton and thermal structure in the upper ocean: consequences of nonuniformity in chlorophyll profile. Journal of Geophysical Research 88(C4):2565-2570.

Lighthart, B. and L.D. Stetzenbach. 1994. Distribution of microbial bioaerosols. In: B. Lighthart and A.J. Mohr (eds.). Atmospheric Microbial Aerosols: Theory and Applications. New York: Chapman and Hall, pp. 68-98.

Lindquist, A. 1968. On fish eggs and larvae in the Skagerrak. Sarsia 34:347.

Lubchenco, J. 1978. Plant species diversity in a marine intertidal community: importance of herbivore food preference and algal competitive abilities. American Naturalist 112:23-39.

Mann, K.H. and J.R.N. Lazier. 1996. Dynamics of Marine Ecosystems: Biological-Physical Interactions in the Oceans. 2nd Edition. London: Blackwell Science.

Marchand, P.J. and D.A. Roach. 1980. Reproductive strategies of pioneering alpine species: seed production, dispersal and germination. Arctic and Alpine Research 12:137-146.

Marshall, N.B. 1966. The Life of Fishes. Cleveland, OH: World Publications.

Mommaerts, J.P., G. Pichot, J. Ozer, Y. Adam and W. Baeyens. 1984. Nitrogen cycling and budget in Belgian coastal waters: North Sea areas with and without river inputs. Rapp. P.-v. Reun. Cons. Int. Explor. Mer. 183:57-69.

Moore, W.S. 1984. Evolutionary ecology of unisexual fishes. In: B.J. Turner (ed.). Evolutionary Genetics of Fishes. New York: Plenum Press.

Morse, A.N.C. 1991. How do planktonic larvae know where to settle? American Scientist 79:154-167.

Morse, D.E. and A.N.C. Morse. 1988. Chemical signals and molecular mechanisms: Learning from larvae. Oceanus 31(3):37-43.

National Research Council (NRC). 1995. Understanding Marine Biodiversity: A Research Agenda for the Nation. Washington, DC: National Academy Press.

Norse, E.A. (ed.). 1993. Global Marine Biological Diversity: A Strategy for Building Conservation into Decision Making. Washington, DC: Island Press.

Odum, E.P. 1971. Fundamentals of Ecology. 3rd Edition. Philadelphia: Saunders.

Owen, O.S. 1975. Man and the ocean. In: Natural Resource Conservation. New York: MacMillian Publishers Co.

Paine, R.T. 1966. Food web complexity and species diversity. American Naturalist 100:65-75.

Pasciak, W.J. and J. Gavis. 1974. Transport limitation of nutrient uptake in phytoplankton. Limnology and Oceanography 19: 881-888.

Pierzynski, G.M., J.T. Sims and G.F. Vance. 1994. Soils and Environmental Quality. London: Lewis Publishers.

Powers, D.A. 1995. New Frontiers in Marine Biotechnology: Opportunities for the 21st Century. In: C.G. Lundin and R.A. Zilinskas (eds.). Marine Biotechnology in the Asian Pacific Region. Washington, DC: The World Bank.

Primack, R.B. 1993. Essentials of Conservation Biology. Sutherland, MA: Sinauer Associates, Inc.

Purcell, E.M. 1977. Life at low Reynolds number. American Journal of Physics 45:3-11.

Randall, J.E. and H.A. Randall. 1963. The spawning and early development of the Atlantic parrotfish, *Sparisoma rubripinne*. Zoologica 48:49.

Rasmusson, E.M. and J.M. Wallace. 1983. Meteorological aspects of the El Niño/southern oscillation. Science 222:1195-1202.

Ray, G.C. 1976. Critical marine habitats. In: Proceedings of an International Conference on Marine Parks and Reserves, Tokyo, Japan 12-14 May 1974. IUCN Publications New Series 37.

Reaka-Kudla, M.L. 1997. The global diversity of coral reefs: A comparison with rain forests. In: M.L. Reaka-Kudla, D.E. Wilson and E.O. Wilson (eds.). Biodiversity II: Understanding and Protecting Our Biological Resources. Washington, DC: Joseph Henry Press.

Reynolds, W.M. and D.A. Thompson. 1974. Responses of young grunion, *Leuresthes sardinia*, to gradients of temperature, light, turbulence and oxygen. Copeia 747.

Ring Group. 1981. Gulf Stream cold-core rings: their physics, chemistry, and biology. Science 212:1091-1100.

Robinson, A.R. 1983. Overview and review of eddy science. In: A. R. Robinson (ed.). Eddies in Marine Science. New York: Springer-Verlag.

Rowell, T.W. and R.W. Trites. 1985. Distribution of larval and juvenile Illex (*Mollusca: Cephalopoda*) in the Blake Plateau region (Northwest Atlantic). Vie Milieu Ser. C 35:149-161.

Royce, W.F. 1984. Introduction to the Practice of Fishery Science. New York: Academic Press.

Royce, W.F., L.S. Smith and A.C. Hartt. 1968. Models of oceanic migrations of Pacific salmon and comments on guidance mechanisms. Fisheries Bulletin 66:441-462.

Ryther, J.H. and E.M. Hulbert. 1960. On winter mixing and the vertical distribution of phytoplankton. Limnology and Oceanography 5:337-338.

Sadovy, Y. and D.Y. Shapiro. 1987. Criteria for the diagnosis of hermaphroditism in fishes. Copeia 1:136-156.

Salthe, S.N. 1969. Reproductive modes and the number and sizes of ova in the urodeles. American Midland Naturalist 81:467-490.

Scheltma, R.S. 1986. On dispersal and planktonic larvae of benthic invertebrates: An eclectic overview and summary of problems. Bulletin of Marine Sciences 39:290-322.

Scheltma, R.S. 1971a. The dispersal of the larvae of shoalwater benthic invertebrate species over long distances by ocean currents. In: D.J. Crisp (ed.). Proceedings of the Fourth European Marine Biology Symposium. Cambridge: Cambridge University Press.

Scheltma, R.S. 1971b. Larval dispersal as a means of genetic exchange between geographically separated populations of shallow-water benthic marine gastropods. Biology Bulletin 140:284-322.

Scheltma, R.S. 1966. Evidence for trans-atlantic transport of gastropod larvae belonging to the genus, *Cymatium*. Deep Sea Research 13:83.

Semtner, A.J. 1995. Modeling ocean circulation. Science 269:1379-1385.

Shanks, A.L. 1985. Behavioral basis of internal-wave induced shoreward transport of megalopae of the crab, *Pachygrapus crassipes*. Mar. Ecol. Prog. Ser. 24:289-295.

Shanks, A.L. 1983. Surface slicks associated with tidally forced internal waves may transport pelagic larvae of benthic invertebrates and fishes shoreward. Mar. Ecol. Prog. Ser. 13:311-315.

Shanks, A.L. and W.G. Wright. 1987. Internal-wave mediated shoreward transport of cyprids, megalopae, and gammarids and correlated longshore differences in the settling rate of intertidal barnacles. Journal of Experimental Marine Biology and Ecology 114:1-13.

Shannon, L.V. 1985. The Benguela ecosystem Part 1. Evolution of the Benguela, Physical features and processes. Oceanography and Marine Biology Annual Review 23:105-182.

Sheldon, J.C. and F.M. Burrows. 1973. The dispersal effectiveness of archenepappus units of selected compositae in steady winds with convection. New Phytol. 72:665-675.

Silvertown, J.W. 1982. Introduction to Plant Population Ecology. London: Longman Group Ltd.

Smayda, T.J. 1970. The suspension and sinking of phytoplankton in the sea. Oceanography and Marine Biology Annual Review 8:353-414.

Soto, C.G., J.F. Leatherland and D.L.G. Noakes. 1992. Gonadal histology in the self-fertilizing hermaphroditic fish, Rivulus marmoratus (Pisces, *Cyprinodontidae*). Canadian Journal of Zoology 70(12):2338-2347.

Steele, J.H. 1991. Marine functional diversity. BioScience 41(7):470-474.

Steele, J.H. 1985. A comparison of terrestrial and marine ecological systems. Nature 313(6001):355-358.

Steele, J.H., S. Carpenter, J. Cohen, P. Dayton and R. Ricklefs. 1989. Comparison of Terrestrial and Marine Ecological Systems. Report of a Workshop, Santa Fe, New Mexico. National Science Foundation Blue Paper (USA).

Stramska, M. and T.D. Dickey. 1993. Phytoplankton blooms and the vertical structure of the upper ocean. Journal of Marine Research 51:819-842.

Strathmann, R.R. 1990. Why life histories evolve differently in the sea. American Zoologist 30: 197-207.

Strickler, J.R. 1984. Sticky water: a selective force in copepod evolution. In: D.G. Meyers and J.R. Strickler (eds.). Trophic Interactions within Aquatic Ecosystems. Washington, DC: American Association for the Advancement of Science.

Sunobe, T. and A. Nakozono. 1993. Sex change in both directions by alteration of social dominance in *Trimma okinawae* (Pisces: *Gobiidae*). Ethology 94(4):339-345.

Sverdup, H.U. 1953. On the conditions for the vernal blooming of phytoplankton. Journal of Cons. Perm. Int. Exp. Mer. 18:287-295.

Swallow, J.C. 1976. Variable currents in mid-ocean. Oceanus 19:18-25.

Taunton-Clark, J. and L.V. Shannon. 1988. Annual and inter-annual variability in the southeast Atlantic during the 20th century. South African Journal of the Marine Sciences 6:97-106.

Taylor, M.M., L. DiMichele and G.J. Leach. 1977. Egg stranding in the life cycle of the mummichog, *Fundulus heteroclitus*. Copeia 397.

Thorsen, G. 1950. Reproductive and larval ecology of marine bottom invertebrates. Biological Review 25:1-45.

Thresher, R.E. 1984. Reproduction in Reef Fishes. Neptune City: T.F.H. Publications.

Todd, C.D. and R.W. Doyle. 1981. Reproductive strategies of marine benthic invertebrates: a settlement-timing hypothesis. Marine Ecology Progress Series 4:75-83.

Tranter, D.J., R.R. Parker and G.R. Creswell. 1980. Are warm-core eddies unproductive? Nature 284:540-542.

Trumble, R.J., O.A. Mathisen and D.W. Stuart. 1981. Seasonal food production and consumption by nekton in the northwest African upwelling system. In: F.A. Richards (ed.). Coastal Upwelling. Washington, DC: American Geophysical Union.

Tunniclife, V. 1992. Hydrothermal vent communities of the deep sea. American Scientist 80:336-349.

Turner, B.J., J.F. Elder Jr., T.F. Laughlin, W.P. Davis and D.S. Taylor. 1992. Extreme clonal diversity and divergence in populations of a selfing hermaphroditic fish. Proceedings of the National Academy of Sciences 89(22):10643-10647.

Walsh, J.J., T.E. Whitledge, W. Esaias, R.L. Smith, S.A. Huntsman, H. Santander and B.R. de Mendiola. 1980. The spawning habitat of the Peruvian anchovy, *Engraulis ringens*. Deep Sea Research 27:1-27.

Ward, A.D. and W.J. Elliot (eds.). 1995. Environmental Hydrology. New York: CRC Lewis Publishers.

Warner, R.R. 1988. Sex changes in fishes: hypotheses, evidence and objections. Env. Biol. Fish 22:81-90.

Warner, R.R. 1984. Mating behavior and hermaphroditism. American Scientist 72:128-162.

Warner, R.R. 1978. The evolution of hermaphroditism and unisexuality in aquatic and terrestrial vertebrates. In: E.S. Reese and F.J. Lighter (eds.). Contrasts in Behavior. New York: Wiley Interscience.

Wooster, W.S., A. Bakun and D.R. Mclain. 1976. The seasonal upwelling cycle along the eastern boundary of the North Atlantic. Journal of Marine Research 34:131-141.

Wootton, R.J. 1992. Fish Ecology. London: Chapman and Hall.

Wootton, R.J. 1979. Energy costs of egg production and environmental determinants of fecundity in teleost fishes. Symp. Zool. Soc. London 44:133-159.

Zeldis, J.R. and J.B. Jillett. 1982. Aggregation of pelagic *Munida gregaria* (*Fabricius*) (*Decapoda, Anomura*) by coastal fronts and internal waves. Journal of Plankton Research 4:839-857.

Chapter 3

Analysis of the Ecological Risks Associated with Genetically Engineered Marine Macroorganisms

JOHN J. GUTRICH[1] AND HOWARD H. WHITEMAN[2]
[1]*Ohio State University and* [2]*Murray State University*

1. INTRODUCTION

In 1987, the U.S. National Academy of Sciences (NAS) reported an "urgent need for the scientific community to provide guidance to both investigators and regulators in evaluating planned introductions of modified organisms from an ecological perspective" (NAS, 1987). As biotechnology and its applications have expanded to include oceanic life forms, the need for ecological analysis from a marine perspective has also increased. This chapter aims to provide specific recommendations and analyses to minimize ecological risks associated with uncontained applications of transgenic marine macroorganisms. To accomplish this goal, we examine marine life history strategies, including unique mechanisms of reproduction and dispersal; discuss the implications for risk assessment of the differences between marine and terrestrial/freshwater systems; propose a model to identify causes of ecological effects associated with introductions of transgenic organisms into the marine environment; and apply the model to outline potential ecological risks posed by the products of marine biotechnology that will have to be considered as regulatory agencies implement formal risk assessment procedures.

2. ECOLOGICAL EFFECTS OF MARINE TRANSGENIC ORGANISMS

Risk is defined as an estimate of the probability of a deleterious event actually occurring (Adam et al., 1993). We define ecological risk, in the case of genetically engineered organisms (GEOs), as the probability that a transgenic organism will negatively affect the natural marine ecosystem into which it is introduced. Ecological effects may include a degradation of the abundance and/or diversity of living organisms in the ecosystem, and/or an alteration of abiotic factors.

The ecological analysis model outlined below is designed to evaluate potential ecological effects associated with introduced transgenic marine organisms. The model sequentially estimates the probability of escape (accidental introduction); the probability that escaped transgenic organisms will establish populations in natural areas; and the probability that established populations of escaped GEOs will cause ecological effects (see Figure 1). Inherent in the model is the need to estimate the rate of establishment and spread of transgenic populations, an essential factor in identifying risk management options when successful establishment is determined to be probable.

Ecological Risk Analysis

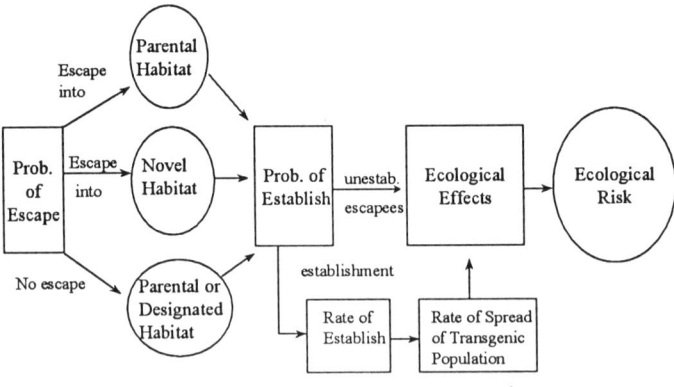

Figure 1. A general risk analysis for assessing the ecological risk of introduced transgenic organisms. This scheme is applicable to both planned and accidental introductions. Planned introductions entail no escape; accidental introductions involve escape into either native (parental) habitat or exotic (novel) habitat.

2.1 Probability of Escape

Escape can be defined as an accidental occurrence in which a transgenic organism is able to utilize novel resources unavailable to the parental species and/or not specified in the proposed introduction. Examples of escape include accidental introduction of an organism beyond an intended biogeographical region, utilization of novel (unplanned) resources by the transgenic organism within the parental or designated habitat of introduction, accidental release of an organism from restrictive facilities into natural ecosystems, and spread of an organism into an exotic habitat beyond the planned habitat of introduction. The probability of a transgenic organism escaping is influenced by the effect transgenic induction has on its ecological constraints. Other critical factors, including life history strategies and environmental characteristics (discussed in detail in Chapter 2), are also considered here as they present additional concerns for risk assessors and managers of field tests of GEOs.

2.1.1 Ecological niche

Ecological niche is an abstract concept that describes all the environmental conditions required for a species to maintain a viable population (Begon et al., 1990; Schoener, 1989). For example, organisms of any given species can maintain a viable and actively growing population only within certain temperature limits (Colwell, 1987). This range of temperature is the ecological niche of the species in one dimension (Begon et al., 1990). Other dimensions include physical/chemical parameters (salinity, pH, dissolved oxygen, etc.) and ecological factors (predators, pathogens, competition, etc.). Of course, an organism is affected by numerous (n) environmental factors; thus the ecological niche can be thought of as an n-hypervolume within which a species can maintain a viable population (Hutchinson, 1957). Most theoretical ecologists are careful to define ecological niche as a property of a particular species, essentially as an extended phenotype (Colwell, 1987; Schoener, 1989). If a marine shellfish species were driven to extinction, the species' resources might just as well be partitioned among zooplankton instead of remaining unutilized until another shellfish species with a similar niche enters the ecosystem. In other words, niches are not compartmentalized spatial entities that become empty, but rather phenotypic responses expressed through competition for resources (Colwell, 1987). A niche cannot be seen, but represents a characteristic of a species. In contrast, habitats are actual places that provide the environmental conditions and resources comprising a niche (Begon et al., 1990).

Ecologists often define niche as either fundamental or realized (Begon et al., 1990). A fundamental niche represents the overall potentialities of a species in the absence of interactions such as competition and predation. Realized niche is the more limited spectrum of conditions and resources that allows a species to maintain a viable population in the presence of such interactions.

Although assessing the probability of escape must be done using a case-by-case approach, a general, more macro-view of transgenic induction and its effects can provide some insight into risk. For example, Table 1, below, describes four scenarios that address the effect of transgenic induction on ecological niche and displays the effect of changes to fundamental and realized niche on the probability of escape and ecological risk. The information provided in Table 1 also serves as a general framework for classifying a transgenic organism as either a native or an exotic species. Transgenic organisms are classified as native species if genetic induction has not altered the fundamental niche, but as exotic if the fundamental niche has been altered.

If it alters the ecological niche of an organism, transgenic induction can directly affect the probability of escape. Transgenic induction could expand the realized and/or fundamental niche of an organism allowing it to obtain novel resources unavailable to the parental species and thus "escape" from the parental ecological niche. For example, transgenic induction can increase the growth rate and body size of an organism. Consequently, the realized niche of the transgenic organism might be expanded if these novel traits decrease predation and competition pressures that define the realized niche of the parent. With an expanded realized niche, the transgenic organism may now possess an advantage over the parent species when competing for limited resources.

Transgenic induction can alter the realized niche without escape as long as the transgenic organism cannot obtain novel resources. The added transgenic property or extended transgenic phenotype may provide no new opportunity to obtain novel resources. For instance, if transgenic induction increases only slightly the body size of a shellfish species, the transgenic organism is most likely still competing with the parent for similar sized prey, thus acting as a native species (Table 1, Scenario 1). Acting as a native species, the relative probability of escape of the transgenic shellfish is low and it poses small ecological risk.

Analysis of the Ecological Risks Associated with Genetically Engineered Marine Macroorganisms

Table 1. Estimates of the probability of escape from the parental habitat based on changes to ecological niche from transgenic induction. Even the slightest alteration of fundamental niche may result in a significant change in the realized niche of transgenic organism as it interacts with novel biotic and abiotic factors. There exists a very high risk that the transgenic organism will escape by obtaining novel resources when transgenic induction has altered the fundamental niche of the organism. In contrast, changes in realized niche (e.g., alterations to competition and predation factors) do not necessarily create changes in fundamental niche of the organism. Case specific ecological analysis would be required to determine which scenario may be applicable to a particular transgenic organism.

Transgenic Induction Scenario	Fundamental Ecological Niche	Realized Ecological Niche	Ecological Risk: Probability of Escape	Classification	Reasoning
1	No change	No significant change	Very low	Native	Realized niche slightly altered; no apparent advantage over parent species.
2	No change	Moderate	Low to medium	Moderately enhanced native	Realized niche moderately altered; slight to moderate advantage over parent species.
3	No change	Significant change	High	Greatly enhanced native	Realized niche significantly altered; substantial advantage over parent species.
4	Any change	Significant change	Very high	Exotic in either parental or native habitat	Fundamental niche altered; ability to obtain novel resources.

Ecological risk increases when transgenic induction enhances significantly a phenotypic trait that increases the opportunity to exploit novel resources. As the change induced by transgenic induction becomes more significant, the larger transgenic shellfish species may realize an advantage over the parental species in obtaining prey (Table 1, Scenarios 2 and 3). If the larger transgenic shellfish species is physiologically able to consume prey that the parent cannot, the transgenic organism may possess the ability to obtain resources unavailable to the parent, increasing the probability of escape and, therefore, ecological risk.

Expansion of fundamental niche implies a significant change in realized niche, resulting in a very high probability of escape and ecological risk (Table 1, Scenario 4). Expansion of the fundamental niche increases the opportunities and potential range of the organism. A transgenic organism that can overcome climatic and other abiotic factors (temperature, pH, salinity, etc.) restricting the parental species may consequently possess the ability to obtain novel resources beyond the parental habitat. For example, genes coding for antifreeze proteins in winter flounder have been transferred, expressed, and inherited in Atlantic salmon (Shears et al., 1991). Expression of antifreeze proteins in Atlantic salmon may allow these GEOs to live in colder habitats than the parent species and, possibly, migrate beyond the current habitat of the parent species. If so, the organism's fundamental niche will have been expanded, increasing ecological risk. Altering the ecological niche by lifting and/or easing ecological constraints can significantly increase the probability of escape.

2.1.2 Ecological constraints

In nature, life history evolution is restricted by phylogenetic and allometric constraints. A phylogenetic constraint refers to the fact that the life history of an organism is limited by historical possibilities available to that organism, and, thus, by its phylogeny. For example, the wandering albatross (*Diomedea exulans*) lays a single egg and has a single brood patch in which it can incubate this egg (Ashmole, 1971). An albatross is unlikely to lay more than one egg because although it might be physiologically and energetically capable of producing multiple eggs, it is constrained by having a single brood patch for incubation. This constraint may result from selection for investment in a single offspring in the ancestors of this species. Thus, such phylogenetic constraints limit the life history variation that can evolve (Begon et al., 1990).

Allometry can be defined as the degree of geometric or physiological similarity among organisms of different size. One such allometric relationship is that fecundity increases with body size. This relationship has been displayed in numerous marine species, including copepods, barnacles, isopods, and mussels (Achituv and Barnes 1978; Daly, 1972; Griffiths, 1977; McLaren, 1965; Platt, 1985; Strong and Daborn, 1979). Organisms are constrained by allometry in that a smaller sized organism will produce fewer eggs than a larger sized individual of the same species.

Life history traits may be influenced by both phylogenetic and allometric constraints. For example, fecundity is constrained phylogenetically to the maximum number of eggs that can be produced utilizing the body that selection favored in past generations. Secondly, fecundity is constrained allometrically by the size of the organism at the current point in time. Selection should favor the optimal combination of life history traits that maximizes reproductive success under these constraints. For instance, fecundity increases with body size, but if an organism delays reproduction until it grows larger it also runs the risk of dying before reproduction (Barnes and Hughes, 1988). Thus, natural selection implicitly addresses the tradeoff between age specific fecundity and mortality, maximizing reproductive success.

It is important to consider phylogenetic and allometric constraints because genetic engineering offers a way of overcoming and/or easing these constraints. The genetic engineering undertaken to date has been focused on inducing or enhancing traits that, as a consequence, may increase the probability of escape and successful establishment of an escapee. For example, the majority of marine transgenic research has focused on creating organisms that are optimal for aquacultural harvesting, possessing characteristics such as large body size, increased growth rate, greater disease resistance, and greater ability to survive adverse environmental conditions relative to the parental species (Zilinskas et al., 1995). Transgenic induction may allow an organism to achieve a phenotype that is unattainable by the parental species, thereby lifting the phylogenetic constraint (Regal, 1994; Tiedje et al., 1989). For example, a GEO could be induced to possess a larger body size (including a larger gape size, larger esophageal basket, larger gills, etc.) that permits the individual to feed on a wider array of larger prey than the parental organism. By lifting the phylogenetic constraint of body size through insertion of novel genes, the transgenic organism can escape from the ecological niche of the parent by obtaining novel resources.

Transgenic induction might also ease the cost of an allometric constraint. For example, Pacific salmon have been transgenically altered to grow four times faster than the parental species (CNI, 1995). As mentioned earlier, if an organism delays reproduction until it reaches a larger size, it runs the risk of dying before reproduction (Barnes and Hughes, 1988). Because the transgenic organism spends less time attaining a larger size and thus spends less time in a critical period in which it may be killed before reproduction (lower mortality rate), the transgenic organism can obtain higher reproductive success. Although transgenic induction alters only growth rate directly, easing the cost of an allometric constraint can result in the significant indirect effect of natural selection favoring larger transgenic organisms over smaller, naturally occurring ones. Examples of transgenic inductions of marine species that serve to lift natural constraints include (1) transgenic induction of Pacific salmon (e.g., coho salmon) to obtain an average mass 11 times larger than that of the parent species (with one extreme example reaching a size 37 times larger) (Devlin et al., 1994; Infofish, 1995); (2) production of transgenic Atlantic salmon that reach sexual maturity in just two years instead of the normal three years (MacKensie, 1996); (3) growth enhancement of Atlantic salmon with the transfer of a chinook salmon GH gene (Du et al., 1992) or fostered by the expression of a mouse methallothionein-human growth hormone fusion gene (Rokkones et al., 1989); (4) gene-coding for antifreeze proteins in winter flounder, which have been transferred, expressed, and inherited in Atlantic salmon (Shears et al., 1991); and (5) growth rate enhancement (up to 40% beyond normal) of the red abalone (Sea Tech., 1995).

2.1.3 The marine environment and life history strategies

Distinctive characteristics of the marine environment and unique marine life histories may cause difficulties in containment and subsequent risk assessment of introduced transgenic organisms not encountered in terrestrial and freshwater introductions (see Table 2, below). Thus, as stated by the Group on Aquatic Biotechnology and Environmental Safety for the Agricultural Biotechnology Research Advisory Committee (ABRAC, 1995), "A thorough review of the life history and environmental requirements of the parental organism is needed in order to determine the potential effects of genetic modification." Characteristics of the marine environment and life history strategies of marine organisms that influence the probability of an escape are addressed in detail in Chapter 2. Here we

Analysis of the Ecological Risks Associated with Genetically Engineered Marine Macroorganisms

Table 2. Comparison of the effects of environmental characteristics and life history strategies on the relative difficulty of containing introduced organisms. Containment is much more difficult, and the probability of escape much greater, in the marine environment than in terrestrial or freshwater environments.

	Terrestrial Environment	Freshwater Environment	Marine Environment
Env. Characteristics			
Medium of Dispersal	Air/transport	Water	Water
Environmental Boundaries	Slightly dynamic	Moderately dynamic	Highly dynamic
Relative Spatial Scale	Segmented (restricted by the seas)	Relatively segmented (restricted by land masses and interfaces with saltwater)	Segmented along coastal areas and by disparate deepsea habitat, yet expansive (70% of earth's surface)
Life History Strategies			
Fecundity	Low	High	Very High
Egg Size	Large	Small to medium	Very small
Planktonic Dispersal	Limited (e.g., wind-blown seeds, bacteria)	Limited (restricted by land masses and saltwater)	Extensive
Planktonic Development	Unavailable	Less common than marine	Very common
Ecological Risk Assessment			
Containment	Manageable	Difficult	Very Difficult
Prob. of Escape	Low	Low to high	Very high

consider how these characteristics may exacerbate risk assessment concerns.

First, the marine environment is conducive to wider and more rapid dispersal of organisms than the terrestrial environment. The oceans provide a scale and continuity of medium that could allow genetically manipulated escapees to influence a geographically broader range of ecosystems. It is conceivable that dispersal patterns for certain marine species may allow transgenic organisms to escape from the intended ecosystem of introduction into waters bordering entirely different countries and/or continents. Further, novel traits from genetic alteration may allow transgenic marine organisms to evade barriers to dispersal that limit the parental species (e.g., nutrient-poor, deepsea habitat that segments coastal populations; see section below entitled "Effect of transgenic induction on ecological constraints"). Thus, the scale and continuity of the oceans require that any risk assessment consider the potential global ecological effects of an accidental introduction, as well as effects on local ecosystems.

2.1.3.1 High fecundity.

High fecundity increases the probability of accidental release since each egg represents an additional entity to be contained. Containment facilities and procedures must be designed to control highly fecund species, which may have egg production rates many orders of magnitude greater than those encountered in terrestrial and freshwater organisms. If those marine organisms most likely to have useful biotechnological applications have high fecundity rates in their natural state, and if these already elevated reproductive potentialities are further enhanced by genetic manipulation, the design of containment facilities will be critical if the potential for escape is to be minimized.

Representatives of environmental organizations have expressed concern about the high fecundity of fish in their discussions of proposals to field test transgenic freshwater fish in aquaculture facilities. In the case of propagation of transgenic carp in experimental outdoor ponds at the Alabama Agricultural Experimental Station, the Environmental Defense Fund recognized the implications of high fecundity on the probability of accidental release. In an effort to decrease any potential risk of environmental effects, they proposed a decrease in the number of larvae utilized (USDA, 1990). Fecundity, on average is higher for marine fish than freshwater fish (Owen, 1975) and, therefore, this concern is even more pertinent when considering ecological risk assessment in the marine environment.

Transgenic inductions of marine organisms so far have focused on organisms useful in aquaculture (Zilinskas et al., 1995). Marine organisms have been genetically engineered to exhibit increased growth rate, disease resistance, and ability to survive adverse environmental conditions. Marine aquaculture may utilize highly fecund marine organisms such as the dolphinfish (*Mahi mahi*) that can spawn naturally every 48 hours, reproduce throughout the year, and produce over 250,000 eggs on each spawning occasion (Nel, 1995). Historically, adequate containment of such highly fecund organisms in aquacultural facilities has been difficult. Courtenay and Williams (1992) note, "In nearly every instance where an aquatic exotic species has been the subject of culture, escape into open waters has occurred...many introduced exotic species have had negative impacts." For example, in August 1995, severe storms caused the release of approximately 150,000 exotic salmon and 50,000 exotic trout from a Chilean aquacultural facility into adjacent waters. The U.S. embassy in Santiago reported that these introduced species could cause negative effects on regional biodiversity (USSD, 1995). Even though such effects have not been noted, experience with containment problems in aquaculture facilities indicates the need for particular attention to proper containment of highly fecund transgenic organisms.

2.1.3.2 Small egg/larva size.

The size of eggs/larvae is a second critical concern in the containment of an introduced transgenic organism. Accidental introductions that have occurred in the past emphasize the importance of egg/larva size in containment. For example, the relatively small size of pink salmon eggs and juveniles allowed nearly 21,000 individuals to be flushed down the drain of a hatchery into a sewer, from which they moved into Lake Superior (Steirer, 1992). In spite of expectations that immature adults could not survive in freshwater, the Great Lakes experienced a population explosion of pink salmon two decades after the introduction (Kwain and Lawrie, 1981; Emery, 1981; ABRAC, 1995). Smaller egg/larva size increases the difficulty with which an organism can be contained and warrants more effective measures to prevent accidental release.

2.1.3.3 Planktonic dispersal and planktotrophic development.

Planktonic dispersal by introduced transgenic marine organisms also has significant implications for containment. In the event of an accidental release, marine organisms possess the ability to travel great distances from the initial point of introduction, in some cases possibly traversing territorial sea boundaries (see Figure 2, below).

Thus, planktonic dispersal creates a unique situation in which the effects of an accidental introduction may occur far from the initial site of application. In contrast, terrestrial and freshwater environments are enclosed by natural boundaries that are more likely to limit the spread of a transgenic organism to the country of introduction. Although marine

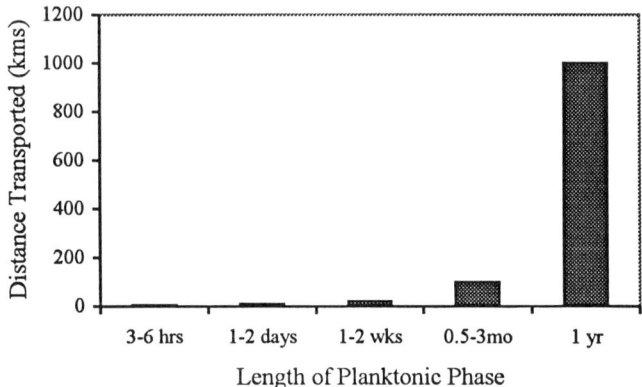

Figure 2. Distance transported as a function of length of marine larval phase. Planktonic dispersal allows an organism to travel great distances from an initial point of introduction (after Crisp, 1978).

organisms are often physiologically restricted to areas along coastlines and in bays, the continuity, scale, and dynamic boundaries of the ocean—in conjunction with a highly utilized long distance dispersal strategy and the novel effects of genetic alteration—may spread transgenic organisms into marine ecosystems of foreign seas. Risk assessment must consider the potential for effects that may occur not only in local ecosystems, but also in ecosystems worldwide.

Since marine eggs/larvae commonly are dispersed as plankton, containment of organisms from planned introductions within a specific biogeographical region, estimates of dispersal area, and recapture of accidentally introduced marine organisms may be difficult. Risk assessment methods, therefore, must incorporate specific currents and other water movements that a marine species may utilize for egg dispersal. Knowledge of dispersal strategies may also provide insight into the biogeographic regions to which an introduced transgenic species is most likely to spread in the event of an escape.

Planktotrophic development of an introduced transgenic organism complicates containment and risk assessment. Planktotrophic development can allow for longer larval periods—because larvae can sustain themselves with available food—and thus may increase dispersal distances (see Figure 2) (Crisp, 1978; Sammarco and Heron, 1994; Young and Eckelbarger, 1994). Utilizing planktotrophic development in conjunction with planktonic dispersal, GEOs could disperse beyond the habitat of parent organisms.

Planktonic dispersal is analogous to the spread of wind-blown seeds on land, but planktonic dispersal in combination with planktotrophic development creates a novel synergistic effect, fostering more effective long distance dispersal unique to the marine environment.

Another facet of this synergistic effect is the correlation between planktotrophy, smaller egg size, and earlier hatching times in nudibranch mollusks. In these organisms, the eggs that become planktotrophic larvae (usually the smallest eggs) hatch the quickest. Time before hatching increases for larger eggs/juveniles produced by lecithotrophic development (utilizing egg yoke) and direct development (Todd and Doyle, 1981). As a result, some marine organisms utilize a reproductive strategy that facilitates long distance dispersal of small eggs that are difficult to contain and also display fast hatching rates. Thus, planktotrophic development may create unique problems for containment and risk assessment of introduced transgenic marine organisms.

2.2 Probability of Successful Establishment

We define "successful establishment" as the production of a viable reproducing population of introduced transgenic organisms in a natural ecosystem and/or production of a population of viable reproducing hybrid organisms possessing the transgene through crossbreeding (population growth in non-target organisms that have received the transgene). When developing data for an ecological risk analysis, it is logical to consider the probability of successful establishment of accidentally introduced transgenic organisms after addressing the probability of escape. Escapees may be organisms initially released in a planned introduction that have accidentally gained access to other ecosystems because of unplanned utilization of novel resources and/or dispersal beyond the initial planned biogeographic region of introduction. Further, GEOs might escape from containment into either the habitat of the parent species or disperse into novel habitats. In general, high mortality rates (as high as 99%) often offset high fecundity levels of marine organisms, thereby reducing opportunities for successful establishment (Barnes and Hughes, 1988). However, in the case of highly fecund marine GEOs it is critical to assess the effect that genetic alteration has on mortality rates. Slight decreases in mortality rate percentages resulting from transgenic induction can lead to large increases in the numbers of surviving offspring, thereby increasing the chance for successful establishment of viable populations. Thus, the probability of successful establishment of escaped transgenic organisms may be correlated with relative fitness conferred by the transgene (e.g., survivability, ability to obtain resources, etc.) and the environmental constraints of the habitat (either parental or novel) into which the organisms are introduced.

In order to estimate the probability of successful establishment of a population of escaped marine GEOs, a number of factors must be taken into account, including (1) effects of transgenic induction on ecological constraints; (2) habitat of introduction; (3) level of competition in the habitat of introduction; (4) number of escapees; (5) modes of reproduction; (6) effectiveness of induced sterility; and (7) active settlement strategies.

2.2.1 Effect of transgenic induction on ecological constraints

Transgenic induction can alter the probability of successful establishment significantly by enhancing or degrading the ability of escaped transgenic organisms to outcompete native species for resources. If the transgenes induced in an organism allow the lifting of phylogenetic and allometric constraints, the transgenic organism might possess the ability to utilize a wider range of life history strategies and an increased number of habitats. The transgenic Atlantic salmon induced with antifreeze proteins may be able to migrate to colder habitats due to the lifting of a phylogenetic constraint. In the case of the red abalone, enhanced growth rate may allow the organism to attain a larger size and higher fecundity than its wild form, which was limited by costly high mortality rates from slower growth (thus, easing an allometric constraint). Further, lifting of both phylogenetic and allometric constraints simultaneously can have additive effects. In the case of the engineered Pacific salmon, the allometric constraint of body size was lifted and the phylogenetically constrained growth rate was enhanced. The transgenic salmon might now possess the ability to obtain extremely high fecundity levels due to larger body size and gain high fecundity at a lower cost because of decreased mortality rates. These factors could increase the probability of successful establishment for transgenic salmon.

2.2.2 Habitat of introduction

The habitat to which a transgenic organism is introduced is a key factor affecting the probability of successful establishment (see Table 3, below). For instance, if a transgenic organism escapes into the habitat of the parent species, the escapee may have to compete with the wild parent for limited

Analysis of the Ecological Risks Associated with Genetically Engineered Marine Macroorganisms

Table 3. Probability of successful establishment in relation to (1) effects of transgenic induction on ecological niche and (2) habitat of introduction. Significant change to the realized niche of an organism results in a high relative chance of establishment in the parental habitat. Change to the fundamental niche may allow the transgenic organism to obtain novel resources unavailable to the parent and thus provides the highest probability of establishment when introduced in the parental habitat. Generally, the likelihood of successful establishment in a novel habitat is low because the introduced organism will most likely encounter environmental factors/resources to which it is not well suited. However, transgenic changes to the fundamental niche may enhance an organism's chance of establishment in a novel habitat. Thus the probability of successful establishment is highly variable and dependent on changes to the fundamental niche and the ecological factors of the novel habitat. Transgenic organisms are classified as exotic if they possess the ability to obtain novel resources and /or are introduced into a novel habitat.

Scenario	Change in Fundamental Niche	Change in Realized Niche	Probability of Establishment in Parental Habitat (A)	Probability of Establishment in Novel Habitat (B)
1	No	No significant change	Low (classified as native for risk assessment)	Low (classified as exotic for risk assessment)
2	No	Moderate change	Medium (classified as native)	Low (classified as exotic)
3	No	Significant change	High (classified as native)	Low (classified as exotic)
4	Any change	Significant change	Very high (classified as exotic)	Highly variable (classified as exotic)

resources (i.e., it acts as a native species). Small changes to the realized niche of the organism (e.g., slight alterations of competition and predation pressures) may offer no new opportunities for it to acquire novel resources, and, therefore, its relative probability of establishment would be low (Table 3, Scenario 1A). The relative probability of successful establishment increases when an organism obtains the ability to utilize new resources within the parental habitat (Table 3, Scenarios 2A and 3A). In this case, transgenic induction allows the organism to procure resources that the parent cannot. In addition, the GEO can take advantage of these resources under abiotic conditions for which the organism is well suited (due to previous natural selection pressures on the parent species from which the transgenic organism was derived). For example, a transgenic Atlantic salmon expressing extraordinary body size may possess a large advantage over the parental species in obtaining prey at temperatures to which both are adapted for survival.

In contrast, if a transgenic salmon were introduced beyond the parental habitat into colder or warmer waters, increased body size would prove to be irrelevant to successful establishment if the transgenic escapee cannot survive in such temperatures (Table 3, Scenario 2B and 3B). When dispersing beyond the parental habitat, the transgenic organism is no longer in direct competition with the parental species nor in an environment to which it is adapted. Under these conditions, the escapee faces changes in factors such as predation pressure, competition, food availability, disease agents and parasites, and geographic restrictions from abiotic factors (such as substrate availability, temperature, salinity, and oceanic currents). Although theoretically the escaped transgenic organism could experience a higher relative fitness from increased access to resources or increased survival, in general the probability of successful establishment of such organisms is rather low. Adaptations of such species are usually not effectively applicable for survival and reproduction in novel habitats (NDBAP, 1988). Thus, probability of successful establishment is generally lower when introduction of an organism occurs in a novel habitat.

In some cases, transgenic induction alters the fundamental niche of an organism. For example, as stated above, genes coding for antifreeze proteins have been transferred into Atlantic salmon (Shears et al., 1991), and these fish may now be suited for colder temperatures, thus expressing a novel ecological niche. However, these transgenic salmon may still face novel biotic factors, such as unique predator/prey relationships, that could diminish survival rates. Further, abiotic factors beyond temperature (e.g., salinity, pH, etc.) may increase mortality (Moyle and Light, 1996). Thus, expansion of the fundamental niche through transgenic induction does not necessarily imply increased successful establishment of the transgenic organism in a novel habitat. The probability of establishment in a novel habitat is highly variable and dependent on case-specific ecological factors (Table 3, Scenario 4B).

The probability of successful establishment is relatively high if transgenic induction significantly alters both the realized and fundamental niche, and the organism escapes into the parental habitat. For example, if Atlantic salmon were transgenically altered to express large body size and produce antifreeze proteins, and if large body size allowed access to novel resources, the organism would possess an expanded fundamental niche within the parental habitat and therefore have a relatively high probability of successful establishment (Table 3, Scenario 4A). A population of extremely large transgenic salmon that can feed on novel prey should have a relatively high chance of establishment in parental waters. In addition, antifreeze proteins would allow the organism to test novel waters. If transgenic salmon migrate beyond the parental habitat and die as a result of detrimental ecological factors, selection will favor establishment within the parental habitat. If the extremely large salmon migrate beyond the parental geographic range and survive, the relative probability of successful establishment is increased. Based on this hypothetical situation, transgenic induction that alters both realized and fundamental niche may create an opportunity for a transgenic organism to expand its habitat, thereby increasing the probability of successful establishment.

2.2.3 Level of competition

The level of competition for limited resources in the habitat to which a transgenic organism has escaped is a key variable affecting the probability of successful establishment. When accidentally introduced into the parental habitat, both the transgenic and parental organisms may search for similar resources (habitats, substrates, nutrients, prey, etc.) and thus the probability of successful establishment of the escapee is related to the level of competition with the parental species. If competition is low, the escapee has a higher chance of survival, thereby increasing the probability of establishment. If competition is high, the chance of survival is low unless transgenic induction increases the organism's competitive ability. In the latter case, it is important to assess the effect of transgenic induction on this variable.

When introduced into a novel habitat, escapees often still compete for limited resources. An escapee may have to compete with native organisms in the new habitat (interspecific competition). For example, denizens of the habitat may feed heavily on prey species that are vital for the survival of the introduced GEO. The exotic transgenic individual must now compete with the natives for food. As before, if competition for food is high, the probability of successful establishment will be low. Under conditions of low competition, the escapee may have a higher probability of becoming successfully established in the novel habitat.

2.2.4 Number of escapees

In dioecious species (those with male and female reproductive organs in separate individuals), the greater the number of escapees, the greater the probability of successful establishment of the transgene in a natural population. This is because the probability of achieving reproductive success in this case increases with the likelihood of finding a mate. Escape of a larger number of transgenic organisms also increases the probability of opportunities for interbreeding with the parent species, thereby making it possible for the transgene to be established in a hybrid form.

Furthermore, when large numbers of individuals escape, the odds are greatly increased that some escapees will find habitats that support their survival and reproduction. For example, when 150,000 exotic salmon and 50,000 exotic trout escaped from a Chilean aquacultural facility, reports expressing concern about damage to biodiversity were immediately released (USSD, 1995). If only a few hundred individuals had escaped, concern for local marine ecosystems may have been much less. (In any case, adverse environmental effects were not observed, suggesting that many factors are at play. Sheer numbers are insufficient to ensure establishment.)

2.2.5 Modes of reproduction

It is important to consider the escapee's reproductive mode when assessing its ability to become successfully established. Upon escape, GEOs exhibiting either simultaneous hermaphroditism or asexual reproduction can achieve reproductive success without having to rely on the probability that another organism of the opposite sex has also escaped, survived, and arrived in the same habitat. Such organisms have an inherently higher probability of successful establishment. Further, in many species of oysters, the organism matures first as a male, slowly becomes a female, quickly becomes a male again, and then alternates sex throughout its life (Royce, 1984). Thus, when the sex ratio of escapees is skewed towards one sex, sequential hermaphroditism provides the opportunity for an introduced organism to change sex, thus increasing the probability of finding a mate.

Certain environmental conditions may also favor reproductive modes that increase the probability of successful establishment. For example, a northern coastal seaweed (*Fucus vesiculosus*) has been shown to retain its gametes in turbulent waters (e.g., when eggs and sperm could be carried away) and release them in tranquil waters, thus resulting in higher fertilization rates (Serrao et al., 1996). Further, two hermaphroditic individuals of the marine shrimp species *Penaeus vannamei* were found among broodstock on a shrimp farm and are believed to have become hermaphroditic as a result of environmental conditions encountered in captivity (Perez-Farfante and Robertson, 1992). Thus, both biotic and abiotic conditions of the habitat of introduction may result in utilization of reproductive strategies that greatly enhance the probability of successful establishment.

2.2.6 Effectiveness of induced sterility

One possibility for decreasing the probability of successful establishment of introduced transgenic organisms is to place a constraint on the organism, such as induced sterility. Induction of permanent infertility would eliminate any ecological risk from successful establishment of escapees. However, the efficacy of introduced sterility in fish and shellfish varies greatly depending on the species, methods (e.g., triploid induction, eyestalk ablation, removal of gonad tissue), protocols for a given method (e.g., specific level, timing and duration of temperature/pressure shock in triploidy induction), and the skill of the operator (ABRAC, 1995).

Artificially induced triploid organisms are usually sterile because their eggs or sperm contain chromosomes that remain unpaired upon fertilization, resulting in death of the embryo. Induction of triploidy in fish and shellfish has been shown to be successful in from 3% to 100% of the organisms altered in a particular trial, with many reports in the 40% to 60% range (Ihssen et al., 1990). Moreover, triploids vary among organisms in terms of development of reproductive structures, reproductive behavior, and presence or absence of gamete production (Hallerman and Kapuscinski, 1993). For example, male triploid shellfish have sometimes been found to produce functional haploid sperm (Allen, 1987); whereas female triploids show very little ovary development (Lincoln and Scott, 1984). Thus, in general, the degree of sterility is considered less complete in triploid male fish and shellfish than in females (Thorgaard and Allen, 1992).

The possibility for restored fertility in triploids has also been discovered in field tests in the York River of the Chesapeake Bay (Blankenship, 1994). In a study of oysters in which sterility had been induced through triploidy, some cells reverted to the diploid state in 20% of the individuals, indicating the possibility for restored fertility to the organisms over time (ABRAC, 1995; S.K. Allen, personal communication, 1996). Another study of the Pacific oyster found that hermaphroditism in an altered, triploid population was markedly higher (29% of individuals expressing this trait) than in a normal, diploid population (1% of individuals expressing hermaphroditism) (Allen and Downing, 1990). Thus, induced sterility is not a guarantee that an organism will be infertile. Triploid organisms may express a variety of reproductive strategies, thereby reducing the effectiveness of induced sterility procedures designed to decrease the probability of successful establishment of escaped GEOs.

2.2.7 Active settlement

Active settlement by marine larvae may greatly enhance the probability of successful establishment upon accidental release. Although deferring settlement and adult development with further larval planktotrophic development cannot ensure survival, it may increase the probability of establishment of an introduced transgenic organism by providing the means for a longer, active search for suitable habitat. Thus, the chances of finding suitable habitat for survival and reproduction are much greater when the larvae are influenced by environmental cues indicating where to settle and provided a longer period of time in which to search for such cues.

2.3 Rate of Establishment and Rate of Spread of Transgenic Populations

Two distinct characteristics of the process of biological invasion include (1) the time required for initial establishment of a viable population of a species at a single location (rate of successful establishment); and (2) expansion over time of the space utilized as the established population grows (rate of expansion of an established population) (Hastings, 1996). Rate differs from probability in that it addresses how quickly an event (e.g., successful establishment or spread of a population) will occur, in contrast to the chance that an event will actually take place. Rate of establishment is important when assessing ecological risk because it serves as a measure of how quickly ecological effects may occur in the initial habitat of introduction as GEO populations become established. Spread is a measure of the spatial magnitude of ecological effects that may occur and represents the geographic expansion of range of introduced organisms. Rate of spread serves as an indicator of how quickly additional marine ecosystems may be susceptible to invasion by escaped transgenic organisms. Upon determination that escaped GEOs can successfully establish a population, estimation/assessment of the rate of establishment and rate of spread of the transgenic population is essential for constructing risk management options (see Chapter 1).

Fecundity may strongly affect the rate at which a transgenic organism will become established. High fecundity can lead to extreme changes in abundance because small changes in the survival rates between spawning and sexual maturity will generate large changes in the absolute abundance of adults in the population (Wootton, 1992). Thus, introduced marine organisms that enter ecosystems with suitable conditions and no naturally occurring predators can quickly establish themselves. For example, the carnivorous American comb jellyfish (*Mnemiopsis leidyi*) was accidentally introduced into the Black Sea in the early 1980s. Within eleven years, it had overrun the habitat, constituting 95% of the wet weight biomass found in the sea (Travis, 1993). Similarly, within 10 years of its introduction into the San Francisco Bay in 1986, the Chinese clam (*Potomocorbula*) reached concentrations of 20,000 per square meter and came to represent greater than 95% of the biomass (Carlton et al., 1990; NRC, 1995).

An increased rate of establishment arising from high fecundity of marine organisms has implications for risk assessment of introduced transgenic organisms. A faster rate of establishment, if conditions are favorable, translates into an increased rate of ecological change than generally seen on land or in freshwater. Further, planktotrophic larvae in nudibranches have been shown to hatch sooner than either lecithotrophic larvae (those utilizing egg yolk from the parent) or directly released juveniles. Thus, ecological effects may occur at a faster rate in the marine environment as a result of the feeding characteristics of planktotrophic larvae. Consequently, if GEOs escape into favorable waters, efforts to remedy the situation (e.g., recapture or eradicate the organisms) may be less successful.

A growing body of literature is emerging in the field of invasion ecology focusing on the spread of introduced species (Grosholz, 1996; Hastings, 1996; Johnson and Carlton, 1996). The rate at which introduced organisms spread has been contrasted for terrestrial and marine systems. Grosholz (1996) compared the rates of spread across terrestrial and marine environments by analyzing observed rates of expansion for well-documented invasions of introduced species. No significant difference was found between the average rate of range expansion in terrestrial and marine habitats. However, the study also revealed that marine species exhibited enormous year-to-year variations in the rate of spread while terrestrial species did not. Grosholz (1996) suggests that the "vagaries in the movements of surface currents may mean that marine species may be transported over long distances in some years, while in other years they may not expand their range at all."

Although preliminary studies indicate that rates of spread in the long-run do not differ across environments, drastic variation in the annual rates of spread of marine species may increase the difficulties faced by risk assessors attempting to predict short-term fluctuations in the abundance and composition of marine species affected by the spread of introduced transgenic organisms. Further ecological studies analyzing the timing of reproductive activities relative to changing oceanographic conditions will be essential for estimating the annual spread of transgenic marine escapees. Extreme short-term variations in the historic spread of introduced marine organisms indicate the possibility for sporadic ecological effects from escaped GEOs that may be difficult to predict and/or assess in the marine environment.

2.4 Ecological Effects of Introduced Transgenic Marine Organisms

Ecological effects of introduced transgenic organisms can be defined as any changes in a given environment (abiotic or biotic) that result from an introduction or escape. Specifically, ecological effects include changes to the composition, abundance, and diversity of living organisms and abiotic factors that constitute the natural environment.

2.4.1 Effects on living organisms

An introduced transgenic organism may directly affect natural ecosystems in a number of ways. For instance, an introduced transgenic organism may influence biotic factors, such as the survival rate of a commercially important and valuable species, through destruction of substrate, consumption of larvae/eggs, consumption of an organism in a lower trophic level that results in a decrease of the prey population for the native species, transportation of disease or another exotic species that may cause similar negative ecological effects, consumption of the adults of the native species directly as prey, and competition with the native species for limited resources (e.g., prey, substrate, refugia, etc.).

Analysis of the Ecological Risks Associated with Genetically Engineered Marine Macroorganisms

Escaped transgenic organisms act as exotic species if they can obtain novel resources unavailable to the parent or when introduced beyond the habitat of the parent. Case studies of ecological effects from introductions of exotic species can serve as indicators of the effects that may occur from introductions of GEOs. Six examples follow. First, the common periwinkle (*Littorina littorea*) that was introduced into North America in the early 1800s established itself as an intertidal snail species and changed the community composition of rocky shores, mudflats, and salt marshes along the northeastern coast of the United States, particularly affecting seaweed populations (Levington, 1995). Today, this organism represents one of the most ecologically important intertidal species of the Atlantic coast (Race, 1982; Carlton, 1985, 1987, 1989; Nicholas and Thompson, 1985; Norse, 1993). Second, arrival of the diatom *Biddulphia sinensis* from the Indo-Pacific region in the early 1900s altered the composition of phytoplankton species in an area of the North Sea by establishing itself as one of the dominant species (Elton, 1958; Hardy, 1956). Third, the Asian eelgrass (*Zostera japonica*) colonizing the northwest coastal shores of the U.S. created vast areas of rooted vegetation where mudflats once existed, leading to distinct changes in the associated infaunal organisms (Carlton, 1989; Norse, 1993; Posey, 1988). Fourth, introduction of the bluestripe snapper (*Lutjanus kasmira*) into the Hawaiian Islands, greatly influenced the relative abundance of native species (Randall, 1987). Fifth, the introduction of the Asian green alga (*Codium fragile tomentosoides*) along the U.S. Atlantic coast has had an extensive impact on local mollusk fisheries (Carlton and Scanlon, 1985). Finally, many Red Sea species introduced through the Suez canal, including fish, shrimp, and jellyfish, have depleted populations of ecologically similar Mediterranean Sea species, as well as some non-similar species (Calman, 1927; Elton, 1958; Fox, 1926; Norse, 1993; Spanier and Galil, 1991).

Introduced transgenic organisms may alter the life history strategies implemented by native species by fostering increased predation or competition for resources. For example, predation pressure applied by herbaceous grazers in the intertidal zones of rocky shores has been shown to influence the implementation of life history strategies in a number of algae (Dirzo, 1984). Observations and experimental removal of herbivores along the shores of New England, Oregon and Washington suggest that algae exhibit an "upright" structure when grazing pressure is low, and a "non-upright fleshy crust" in times of high grazing pressure (Lubchenco and Cubit, 1980; Slocum, 1980). Strong competition between marine organisms has also been shown to affect an organism's life history strategies. Menge (1974, 1975) studied two species of intertidal seastars, (*Leptasterias hexactis* and *Pisaster ochraceus*), and found that when living under competitive conditions in the presence of *Pisaster*, *Leptasterias* grew to a smaller size and brooded its eggs. If *Pisaster* was removed, *Leptasterias* grew larger and utilized a different life history strategy: the release of pelagic eggs. Thus, another rather distinctive effect of an introduced transgenic organism is the effect it may have on life history strategies utilized by native organisms.

Ecological effects can foster feedback loops that exacerbate the magnitude of environmental alterations by causing "cascading" effects. An introduced transgenic organism may decrease the abundance of a single species, which, in turn, can significantly alter the composition of a community. For example, removal of the starfish *Pisaster ochraceus* from waters off the coast of Washington allowed other organisms to invade or overgrow the habitat (Paine, 1966; Stone, 1995). Similar cascading ecological effects are possible when an introduced transgenic organism consumes large quantities of the larvae or eggs of a native species. An introduced transgenic organism feeding on the larvae of a native crustacean directly decreases the number of individuals in the crustacean population. Decreasing the numbers of a particular crustacean may correspondingly increase the abundance of zooplankton and phytoplankton populations on which they feed. In turn, the larger populations of phytoplankton and zooplankton may serve as an increased food source for larvae of other macro-organisms, such as fish, causing subsequent increases in their population levels. Thus, ecological effects from increased predation on a particular species can affect a variety of different organisms that are interconnected in the food web that may further magnify levels of ecological change.

2.4.2 Effects on abiotic factors

Introduced transgenic organisms can also influence abiotic factors. Waste material released by an extremely large transgenic salmon may influence nutrient concentrations in the natural ecosystem. Such alterations could lead to blooms of microorganisms that depend on nutrients such as nitrogen and phosphorus released from the organically bound state in the escapee's excrement (Ryther and Dunstan, 1971; Valiela, 1991). In turn, ecological effects from alteration of abiotic factors may be quite extensive. Blooms of phytoplankton resulting from nutrient fluxes can block sunlight from reaching macroalgae at lower depths and thus deplete marine plant populations. Further, many phytoplankton species release toxins that inhibit the growth of competing species (Norse, 1993). Blooms of the dinoflagellate *Chatonella* result in toxic red tides that have caused massive fish kills in Florida (Halverson and Martin, 1980). The dinoflagellate *Gonyaulax* has caused mass mortality of wedge clams (*Donax serra*) in South Africa (Villier, 1979). In Canada, the dinoflagellate toxin of *Gonyaulax* has been transferred through the food web killing Atlantic herring (*Clupea harengus harengus*) that feed on contaminated herbivorous zooplankton (White, 1980). Decomposition of an extensive number of dead GEOs could also serve to alter local nutrient levels and the abundance of microorganisms that feed on detritus (Fenchel, 1977). Thus, ecological effects causing alterations of abiotic factors may be as important as those arising from biotic changes.

To summarize, direct and indirect alterations of abiotic and biotic factors in a habitat of introduction can induce ecological effects. Also, it is important to recognize that ecological effects may be observed even if an escapee does not become successfully established. Altered nutrient levels and predator/prey relationships in a habitat of introduction may occur within the lifetime of an escapee even if the individual does not successfully reproduce. Therefore, risk assessment protocols must consider these effects as well.

Ecological effects arising from successfully established populations are considered to be more critical because effects caused by unestablished escapees tend to be short-lived. Ecological effects caused by unestablished escapees would most likely be of relatively short duration because direct alteration to abiotic and biotic factors will end with the death and full decomposition of the escapee. In contrast, an established population of transgenic escapees, or hybrids possessing the transgene, could continue to influence abiotic and biotic factors of an ecosystem for a longer period of time. Case-by-case analysis is required to determine specific ecological effects that may occur from an introduction and the time period over which these effects may persist.

3. ESTIMATING THE PROBABILITY OF ECOLOGICAL EFFECTS

To this point, we have presented a general scheme that provides a sequential methodology for assessing the likelihood of ecological effects yielding results that can be used by risk assessors in formal protocols relating to hazard and exposure (Figure 1). Upon collection of case-specific ecological information, the likelihood of ecological effects can be estimated by assessing (1) the probability of escape; (2) the probability of successful establishment; (3) the rate of establishment; (4) the rate of spread of the population; and (5) potential ecological effects. Application of this general risk assessment scheme can be applied on a macro-scale and thus can provide a broad measure of relative ecological risk. We have analyzed each of the five proposed components of ecological risk for terrestrial, freshwater, and marine environments (see Table 4, below). The analysis is general and serves as an indicator of similarity or disparity of ecological risk across environments. Assessing the ecological risk of introducing a particular organism will require an extensive, case-specific analysis of ecological factors.

Table 4. Estimates of relative ecological risk derived from (1) probability of escape, (2) probability of successful establishment (including rate of establishment and population spread upon establishment), and (3) ecological effects for the terrestrial, freshwater, and marine environments. Ecological risk is estimated to be high in the marine environment, relative to land and freshwater environments, because (1) the probability of escape is very high, (2) the probability of successful establishment is moderate, (3) the rate of establishment is high, (4) the rate of population spread is moderate, and (5) the potential for further ecological effects is high.

Environment	Prob. of Escape	Prob. Of Establishment	Rate of Establishment	Rate of Spread	Ecological Effects	Ecological Risk
Terrestrial	Low	Medium	Low	Medium	Low	Low
Freshwater	Medium	Medium	Medium	Medium	Medium	Medium
Marine	Very high	Medium	High	Medium	High	High

We estimate that the relative probability of escape in the marine environment is high. Marine organisms are often highly fecund, produce relatively small eggs/larvae, employ planktotrophic development, and utilize planktonic dispersal, making containment difficult. Further, the size, continuity, and dynamic boundaries of the seas contribute to a relatively high chance of escape, combined with the demonstrated ability of marine larvae to traverse great distances (see Chapter 2).

The probability of successful establishment is considered in either a parental or exotic habitat. Assessing the probability of successful establishment of transgenic organisms requires analysis beyond parental mortality rates. In marine systems, one may falsely conclude that as a result of generally high mortality rates for the parental species the probability of successful establishment of transgenic marine organisms will be low. In actuality, the probability of successful establishment of transgenic organisms for their respective environments is highly dependent on transgenic induction, habitat of introduction, and competition for limited resources (Table 3). The large range and high variability of these factors may greatly enhance or decrease mortality rates of an escapee in all three environments. Thus, we estimate the overall probability of establishment, as the average of these effects, to be moderate.

It is important to consider the high ecological risk from outliers (e.g., extreme or unusual cases) even though the average chance of establishment may be moderate. Introductions of significantly enhanced GEOs into the parental habitat may create high ecological risk (Table 3, Scenarios 3A and 4A). Further, marine life history strategies, such as active settlement, may enhance the probability of establishment. Thus, although there is no distinctive difference across environments when assessing the chances of successful establishment, it is relevant to conduct case-specific analysis of ecological factors to identify any highly risky scenario.

Rate of establishment and rate of spread are relevant if successful establishment of a transgene in the natural biota is to occur. We estimate that GEOs introduced into the ocean environment may have a relatively high rate of establishment because of the relatively high fecundity of many marine organisms. In contrast, the rates of establishment on land and in freshwater are relatively lower because there exist lower average levels of fecundity among these organisms. Preliminary analysis has yielded results showing no significant difference between average rates of range expansion of species across terrestrial and marine systems. Since insufficient evidence exists to definitively indicate a disparity across environments for rates of spread, we estimate a moderate relative risk for each environment. However, results of early studies suggest large annual variations in range expansion for marine species and emphasize the importance for risk assessment of the sporadic, short-term rapid rates of spread that may be more likely to occur in the marine environment.

Ecological effects can foster feedback loops that increase the probability of further ecological effects, thereby increasing ecological risk. Our analysis suggests that ecological change will be more likely in marine systems, and that there exists a relatively high ecological risk that further cascading effects will occur.

Overall, we estimate ecological risk posed by introduced GEOs in the marine environment to be relatively high compared to terrestrial and freshwater environments. The factors contributing to this assessment are that the probability of escape of introduced organisms is very high, the probability of successful establishment is moderate, the rate of establishment of escapee populations is likely to be high, the rate of spread is likely to be moderate, and the potential for further ecological effects in marine systems is high compared to freshwater or terrestrial environments.

4. CONCLUSION

On a macro scale, the questions to be asked when evaluating ecological risk are similar for terrestrial, freshwater, and marine environments. However, the unique characteristics and greater complexity of oceanic ecosystems increase both the likelihood that introduced organisms will escape and the difficulty of accurately estimating the potential ecological effects. Identification of biogeographical ranges and dispersal patterns of organisms that can be performed with relative ease on land and in freshwater systems are more difficult in the marine environment. Furthermore, effects in marine ecosystems may occur at a faster rate. As a result, the margin for error in ecological risk assessment is narrower for the marine environment compared to terrestrial and freshwater ecosystems, and more detailed analyses are warranted when considering introductions of transgenic organisms into marine ecosystems. In some cases, a lack of understanding of marine abiotic and biotic variables may require further ecological research before such an introduction can be approved.

In summary, the analysis presented here suggests that the ecological risk associated with GEOs is higher in the marine environment than in freshwater or terrestrial systems. This heightened risk must be considered when assessing proposals to introduce genetically engineered organisms into the open oceans.

REFERENCES

Achituv, Y. and H. Barnes. 1978. Some observations on *Tetraclita squamosa rufotincta*. J. Exp. Mar. Biol. Ecol. 31:315-324.

Adam, K.D., C.M. King and W.H. Kohler. 1993. Potential effects of escaped transgenic animals: lessons from past biological invasions. In: K. Wohrmann and J. Tomiuk (eds.). Transgenic Organisms: Risk Assessment of Deliberate Release. Boston: Birkhauser Verlag.

Agricultural Biotechnology Research Advisory Committee (ABRAC). 1995. Performance Standards for Safely Conducting Research with Genetically Modified Fish and Shellfish. Document No. 95-04. Washington: United States Department of Agriculture.

Allen, S.K. Jr. 1996. Personal communication: concerning risk assessment and triploidy in the Pacific oyster, *Crassostrea gigas*, held at the Haskins Research Laboratory, Rutgers University. September 17th.

Allen, S.K. Jr. 1987. Reproductive sterility of triploid shellfish and fish. Ph.D. Dissertation, University of Washington, Seattle.

Allen, S.K. Jr. and S.L. Downing. 1990. Performance of triploid Pacific oysters, *Crassostrea gigas*: gametogenesis. Canadian Journal of Fisheries and Aquatic Sciences 47:1213-1222.

Ashmole, N.P. 1971. Sea bird ecology and the marine environment. In: D.S. Farner and J.R. King (eds.). Avian Biology. New York: Academic Press.

Barnes, R.S.K. and R.N. Hughes. 1988. An Introduction to Marine Ecology. 2nd Edition. Boston: Blackwell Scientific Publications.

Begon, M., J.L. Harper and C.R. Townsend. 1990. 2nd Edition. Ecology: Individuals, Populations and Communities. Boston: Blackwell Scientific Publications.

Blankenship, K. 1994. Experiment with Japanese oyster ends abruptly: oysters thought to be sterile found capable of reproducing. Bay Journal 4(5):1-4. Baltimore, MD: Alliance for Chesapeake Bay.

Calman, W.T. 1927. Zoological results of the Cambridge Expedition to the Suez Canal, 1924. XIII Report on the Crustacea *Decapoda* (*Brachyura*). Trans. Zool. Soc. Lond. 22:211-219.

Carlton, J.T. 1985. Transoceanic and interoceanic dispersal of coastal marine organisms: the biology of ballast water. Oceanography and Marine Biology, An Annual Review 23:313-371.

Carlton, J.T. 1987. Mechanisms and patterns of transoceanic marine biological invasions in the Pacific Ocean. Bulletin of Marine Science 41:467-499.

Carlton, J.T. 1989. Man's role in changing the face of the ocean: biological invasions and implications for conservation of near-shore environments. Conservation Biology 3(3):265-273.

Carlton, J.T. and J.A. Scanlon. 1985. Progression and dispersal of an introduced alga: *Codium fragile* sp. *tomentosoides* (*Chlorophyta*) on the Atlantic coast of North America. Botanica Marina 28:155-165.

Carlton, J.T., J.K. Thompson, L.E. Schemel and F.H. Nichols. 1990. Remarkable invasion of San Francisco Bay (California, USA) by the Asian clam *Potomocorbula amurensis*. Mar. Ecol. Prog. Ser. 66:81-94.

Colwell, R.K. 1987. Ecology and biotechnology: expectations and outliers. In: J. Fiskel and V. T. Covello (eds.). Safety Assurance for Environmental Introductions of Genetically-Engineered Organisms. New York: Springer-Verlag.

Community Nutrition Institute (CNI) and the Biotechnology Working Group. 1995. Transgenic Fish: The Next Threat To Marine Biodiversity. Washington: Community Nutrition Institute.

Courtenay, W.R. Jr. and J.D. Williams. 1992. Dispersal of exotic species from aquaculture facilities, with emphasis on freshwater fishes. In: A. Rosenfield and R. Mann (eds.). Dispersal of Living Organisms into Aquatic Ecosystems. College Park, MD: Maryland Sea Grant College.

Crisp, D.J. 1978. Genetic consequences of different reproductive strategies in marine invertebrates. In: B. Battaglia and J. A. Beardmore (eds.). Marine Organisms: Genetics, Ecology and Evolution. New York: Plenum Press.

Daly, J.M. 1972. The maturation and breeding biology of *Harmothoe imbricata* (*Polychaeta: Polynoidae*). Marine Biology 12:53-66.

Devlin, R.H., T.Y. Yesaki, C.A. Blagi, E.M. Donaldson, P. Swanson and C. Woon-Khlong. 1994. Extraordinary salmon growth. Nature 371:209-210.

Dirzo, R. 1984. Herbivory: a phytocentric overview. In: R. Dirzo and J. Sarukhan (eds.). Perspectives on Plant Population Ecology. Sunderland, MA: Sinauer Associates, Inc.

Du, S.J., Z. Gong, G.L. Fletcher, M.A. Shears, M.J. King, D.R. Idler and C.L. Hew. 1992. Growth enhancement in transgenic Atlantic salmon by the use of an "all fish" chimeric growth hormone gene construct. Bio/Technology 10:176-181.

Elton, C.S. 1958. Changes in the sea. In: The Ecology of Invasions By Animals and Plants. London: Methuen and Co. Ltd.

Emery, L. 1981. Range extension of pink salmon into the lower Great Lakes. Fisheries 6(2):7-10.

Fenchel, T.C. 1977. Aspects of the decomposition of seagrasses. In: C.P. McRoy and C. Helfferich (eds.). Seagrass Ecosystems: A Scientific Perspective. New York: Dekker.

Fox, H.M. 1926. Zoological results of the Cambridge Expedition to the Suez Canal, 1924. I. General part. Trans. Zool. Soc. Lond. 22:1-64.

Gittleman, J.L. 1985. Carnivore body size: ecological and taxonomic correlates. Oecologia 67:540-544.

Griffiths, R.J. 1977. Reproductive cycles in littoral populations of *Choromytilus meridionalis* (Kr.) and *Aulacomya ater* (*Molina*) with a quantitative assessment of gamete production in the former. J. Exp. Mar. Biol. Ecol. 30:53-71.

Grosholz, E.D. 1996. Contrasting rates of spread for introduced species in terrestrial and marine systems. Ecology 77(6):1680-1686.

Hallerman, E.M. and A.R. Kapuscinski. 1993. Potential impacts of transgenic and genetically manipulated fish on natural populations: addressing the uncertainties through field testing. In: J. Cloud (ed.). Conservation Genetics of Salmonid Fishes. New York: Plenum Press.

Halverson, M. and D.F. Martin. 1980. Studies of cytoclysis of *Chattonella subsalsa*. Florida Science 43(1):35.

Hardy, A.C. 1956. The Open Sea. Its Natural History: The World of Plankton. London.

Harris, G.P. 1985. The answer lies in the nesting behavior. Freshwater Biology 15:375-380.

Hastings, A. 1996. Models of spatial spread: Is the theory complete? Ecology 77(6):1675-1679.

Hutchinson, G.E. 1957. The multivariate niche. Cold Spring Harbor Symposium on Quantitative Biology 22:415-427.
Ihssen, P.E. et al. 1990. Ploidy manipulation in fishes: cytogenetic and fisheries applications. Transactions of the American Fisheries Society 119:698-717.
INFOFISH International. 1995. Heavier fish through genetic engineering. INFOFISH International 2:72.
Johnson, L.E. and J.T. Carlton. 1996. Post-establishment spread in large-scale invasions: dispersal mechanisms of the zebra mussel *Dreissena Polymorpha*. Ecology 77(6):1686-1690.
Kwain, W. and A.H. Lawrie. 1981. Pink salmon in the Great Lakes. Fisheries (Bethesda) 6(2):2-6.
Lawton, J.H. 1989. Food webs. In: J.M. Cherrett (ed.). Ecological Concepts: The Contribution of Ecology to an Understanding of the Natural World. Boston: Blackwell Scientific Publications.
Levington, J.S. 1995. Marine Biology: Function, Biodiversity, Ecology. New York: Oxford University Press.
Lincoln, R.F. and A.P. Scott. 1984. Sexual maturation in triploid rainbow trout, *Salmo gairdneri*. J. Fish Biol. 25:385-392.
Lubchenco, J. and J. Cubit. 1980. Heteromorphic life histories of certain marine algae as adaptations to herbivory. Ecology 61:676-687.
MacKensie, D. 1996. Can we make supersalmon safe? New Scientist Jan. 27:14-15.
McLaren, I.A. 1965. Some relationships between temperature and egg size, body size, development rate and fecundity, of the copepod *Pseudocalanus*. Limnol. Oceanogr. 10:528-538.
Menge, B.A. 1975. Brood or broadcast? The adaptive significance of different reproductive strategies in the two intertidal seastars *Leptasterias hexactis* and *Pisaster ochraceus*. Marine Biology 31:87-100.
Menge, B.A. 1974. Effect of wave action and competition on brooding and reproductive effort in the seastar, *Leptasterias hexactis*. Ecology 55:84-93.
Moyle, P.B. and T. Light. 1996. Fish invasions in California: Do abiotic factors determine success? Ecology 77(6):1666-1670.
National Academy of Sciences (NAS). 1987. Introduction of Recombinant DNA-Engineered Organisms into the Environment: Key Issues. Washington: National Academy of Sciences Press.
National Research Council (NRC). 1995. Understanding Marine Biodiversity: A Research Agenda for the Nation. Committee on Biological Diversity in Marine Systems. Washington: National Academy Press.
Nel, S. 1995. Mahi mahi culture goes commercial. INFOFISH International 4:25-29.
New Developments in Biotechnology Advisory Panel. 1988. Field-Testing Engineered Organisms: Genetic and Ecological Issues. Lancaster, PA: Technomic Publishing Co., Inc.
Nicholas, F.H. and J.K. Thompson. 1985. Persistence of an introduced mudflat community in South San Francisco Bay, California. Marine Ecology Progress Series 24:83-97.
Norse, E.A. 1993. Global Marine Biological Diversity: A Strategy for Building Conservation into Decision Making. Washington: Island Press.

Owen, O.S. 1975. Man and the ocean. In: Natural Resource Conservation. New York: MacMillian Publishers Co.

Paine, R.T. 1966. Food web complexity and species diversity. American Naturalist 100:65-75.

Perez-Farfante, I. and L. Robertson. 1992. Hermaphroditism in the penaeid shrimp *Penaeus vannamei* (*Crustacea: Decapoda: Penaeidae*). Aquaculture 103(3-4):367-376.

Platt, T. 1985. Structure of marine ecosystems: its allometric basis. Canadian Bulletin of Fisheries and Aquatic Sciences 213:55-64.

Posey, M. 1988. Community changes associated with the spread of an introduced seagrass *Zostera japonica*. Ecology 69:974-983.

Race, M.S. 1982. Competitive displacement and predation between introduced and native mud snails. Oecologia 41:337-347.

Randall, J.E. 1987. Introduction of marine fishes to the Hawaiian Islands. Bulletin of Marine Science 41:490-502.

Regal, P.J. 1994. Scientific principles for ecologically based risk assessment of transgenic organisms. Molecular Ecology 3:5-13.

Rokkones, E., H. Alestrom, H. Skjervold and K.M. Gautvik. 1989. Microinjection and expression of a mouse metallothionein human growth hormone fusion gene in fertilized salmonid eggs. Journal of Comparative Physiology 158:751-758.

Royce, W.F. 1984. Introduction to the Practice of Fishery Science. New York: Academic Press.

Ryther, J.H. and W.M. Dunstan. 1971. Nitrogen, phosphorus and eutrophication in the coastal marine environment. Science 171:3975.

Sammarco, P.W. and M.L. Heron. 1994. The Bio-Physics of Marine Larvae Dispersal. Washington: American Geophysical Union.

Schoener, T.W. 1989. The ecological niche. In: J.M. Cherrett (ed.). Ecological Concepts: The Contribution of Ecology to an Understanding of the Natural World. Boston: Blackwell Scientific Publications.

Sea Technology. 1995. Growth rate of red abalone increased through biotechnology. Sea Technology 36:(8).

Serrao, E.A., G. Pearson, L. Kautsky and S.H. Brawley. 1996. Successful external fertilization in turbulent environments. Proc. Nat. Acad. Sci. USA. 93:5286-5290.

Shears, M.A., G.L. Fletcher, C.L. Hew, S. Gauthier and P. L. Davies. 1991. Transfer, expression and stable inheritance of antifreeze proteins in Atlantic salmon. Molecular Marine Biology and Biotechnology 1(1):58-63.

Slocum, J.C. 1980. Differential susceptibility to grazers in two phases of an intertidal alga: advantages of heteromorphic generations. J. Exp. Mar. Biol. Ecol. 46:99-110.

Spanier, E. and B.S. Galil. 1991. Lessepsian migration: a continuous biogeographical process. Endeavor, New Series 15(3):102-106.

Steirer, W.F. 1992. Historical perspectives on exotic species. In: M.R. DeVoe (ed.). Introductions and Transfers of Marine Species: Achieving a Balance Between Economic Development and Resource Protection. Hilton Head Island, SC: South Carolina Sea Grant Consortium.

Stone, R. 1995. Taking a new look at life through a functional lens. Science 269:316-317.

Strong, K.W. and G.R. Daborn. 1979. Growth and energy utilization of the intertidal isopod *Idotea baltica* (*Pallus*) (*Crustacea: Isopoda*). J. Exp. Mar. Biol. Ecol. 41:101-123.

Thorgarrd, G.H. and S.K. Allen. 1992. Environmental impacts of inbred, hybrid and polyploid aquatic species. In: A. Rosenfield and R. Mann (eds.). Dispersal of Living Organisms into Aquatic Ecosystems. College Park, MD: Maryland Sea Grant College.

Tiedje, J.M., R.K. Colwell, Y.L. Grossman, R.E. Hodson, R.E. Lenski, R.N. Mack and P.J. Regal. 1989. The planned introduction of genetically engineered organisms: ecological considerations and recommendations. Ecology 70:298-315.

Todd, C.D. and R.W. Doyle. 1981. Reproductive strategies of marine benthic invertebrates: a settlement-timing hypothesis. Marine Ecology Progress Series 4:75-83.

Travis, J. 1993. Invader threatens Black, Azov seas. Science 262:1366-1367.

United States Department of Agriculture (USDA). 1990. Finding of No Significant Impact: Environmental Assessment of Research on Transgenic Carp in Confined Outdoor Ponds to be Conducted at the Alabama Agricultural Experiment Station (AAES) Auburn University, Auburn, Alabama. Washington: USDA.

United States State Department (USSD). 1995. Big storms liberate 200,000 farmed salmon and trout. News release from the American Embassy in Santiago to the Secretary of State, Washington, DC.

Valiela, I. 1991. Ecology of water columns. In: R.S. Barnes and K.H. Mann (eds.). Fundamentals of Aquatic Ecology. 2nd edition. Boston: Blackwell Scientific Publications.

Villier, G. 1979. Recovery of a population of white mussels *Donax serra* at Elands Bay, South Africa, following a mass mortality. Fisheries Bulletin - South Africa - Sea Fisheries Branch 12:69-74.

Warner, R.R. 1988. Sex changes in fishes: hypotheses, evidence and objections. Env. Biol. Fish 22:81-90.

Warner, R.R. 1984. Mating behavior and hermaphroditism. Amer. Sci. 72:128-162.

White, A.W. 1980. Recurrence of kills of Atlantic herring (*Clupea harengus harengus*) caused by dinoflagellate toxins transferred through herbivorous zooplankton. Canadian Journal of Fisheries and Aquatic Science 37(2):2262-2265.

Wootton, R.J. 1992. Fish Ecology. London: Chapman and Hall.

Young, C.M. and K.J. Eckelbarger (eds.). 1994. Reproduction, Larval Biology and Recruitment of the Deep-Sea Benthos. New York: Columbia University Press.

Zilinskas, R.A., R.R. Colwell, D.W. Lipton and R.T. Hill. 1995. The Global Challenge of Marine Biotechnology: A Status Report on the United States, Japan, Australia and Norway. College Park, MD: Maryland Sea Grant College.

Zilinskas, R.A. and C.G. Lundin. 1993. Marine Biotechnology and Developing Countries. World Bank Discussion Paper 210. Washington: World Bank.

Chapter 4

Analysis of the Ecological Risks Associated with Genetically Engineered Marine Microorganisms

RAYMOND A. ZILINSKAS
University of Maryland Biotechnology Institute

1. INTRODUCTION

Research is proceeding throughout the world to develop genetically engineered marine microorganisms (GEMMs) for a wide variety of environmental and industrial uses. It is possible therefore that the first uncontained applications of transgenic organisms in the marine environment will involve microorganisms rather than fish or shellfish. As products are readied for experimental or commercial trials, regulators will be required to assess the potential ecological risks posed by these organisms as part of the approval process. This will present a challenge similar to that faced by risk assessors and regulators in 1984 and 1985 when the transgenic microorganism *Pseudomonas syringae*, also known as ice-minus, was developed for agricultural use in the terrestrial environment. At that time we knew almost nothing about interactions between wild and transgenic organisms, and methods for detecting and monitoring gene flow between transgenic and wild organisms were rudimentary. Similarly, little is known today about how GEMMs are likely to interact with wild marine micro- and macroorganisms or how gene flow occurs in the marine environment (Edwards, 1993). In fact, less is now known about marine ecology than was known about terrestrial ecosystem function in the mid-1980s. As a result, the level of scientific uncertainty, and the challenge to risk assessment, associated with transgenic marine microorganisms is greater than was the case with terrestrial products 15 years ago.

In order to lay the groundwork for evaluating ecological risks posed by GEMMs that may be released into the ocean environment, this chapter considers (1) types of marine microorganisms that are candidates for genetic engineering research; (2) types of applications for which GEMMs are likely to be developed; and (3) risk assessment of uncontained applications of GEMMs. The chapter concludes with a discussion of the barriers preventing application of GEMMs in the marine environment for the present, and what may be done to surmount them.

2. TYPES OF MARINE MICROORGANISMS

Many varieties of microscopic animals and plants inhabit the marine environment, including archaea, bacteria, cyanobacteria, fungi, meiofauna, microalgae, protozoa, and viruses (Zilinskas et al., 1995). Some of these ultimately may be subjected to genetic engineering to create products useful for scientific research or commercial applications.

2.1 Archaea

Archaea resemble bacteria in size and morphology but differ sufficiently in genetic makeup to have been placed in a different domain (Hugenholtz and Pace, 1996). Archaea were first discovered only in 1977. Once investigators were aware of their existence, other archaea species were detected and described in rapid succession. The earliest archaea were recovered from extreme environments—habitats that are anaerobic or characterized by severe conditions such as high temperature, low pH, high salt concentration, etc. An example of an archaea extremophile is *Methanococcus jannaschii*, an organism that lives several thousand meters below the ocean surface near deep sea hot vents at temperatures approaching 100° Celsius, replicates using only inorganic chemicals, and produces methane as a byproduct (Anonymous, 1996a). More recent investigations, however, have found that archaea make up a significant proportion of marine plankton found in conventional environments (Massana et al., 1997). As research with archaea proceeds, they probably will be found to be ubiquitous in nature (Olsen, 1994).

2.2 Bacteria

Systematic bacteriology traces to the work of Louis Pasteur, Robert Koch, and others in the early 1870s. Marine microbiology, however, is relatively undeveloped when compared to microbiology of terrestrial environments. Work by pioneering marine microbiologists in the 1940s and 1950s, such as W. D. Rosenfeld and C. Zobell, did much to clarify the huge numbers of microorganisms inhabiting the oceans and their tremendous diversity. Nevertheless, we still know little of what there is to know about marine microorganisms and their diversity. In 1965, R. MacLeod estimated that microbiologists could culture only about 1% of marine bacteria observed in seawater by microscopy (MacLeod, 1965); almost 30 years later even this figure is considered overly optimistic. The revised estimate is that less than 0.1% of marine bacterial species have been cultured, although some samples may yield higher percentages (Jensen and Fenical, 1994; DeLong, 1997). The revised figure reflects better recognition of the microbial diversity of the marine environment (Staley et al., 1997). The inability of scientists to culture most marine bacteria probably stems from the fact that we have not discovered the precise nutrients and physical conditions required by marine bacteria for growth and multiplication; i.e., the environment appropriate for their multiplication has not yet been replicated in the laboratory.

Significant progress has been made in circumventing the barriers to classical culturing methodology over the last few years. Through the application of molecular phylogenic analysis, many bacterial species and genera have been identified and characterized (Relman and Persing, 1996; Hugenholtz and Pace, 1996; Pace, 1997). Ribosomal RNA sequences, mainly 16S, but also 5S, are being employed to investigate bacteria and plankton abundance in lakes, coastal areas, and open oceans (Schmidt et al., 1991; Thomas, 1996). Despite these advances, much remains to be done in marine microbiology before we are anywhere close to achieving the same level of knowledge of marine bacteria as terrestrial bacteria.

Three general types of bacteria can be found in the oceans. The first, transported bacteria, includes terrestrial and freshwater bacteria accidentally dislodged from their usual terrestrial or freshwater niches into the marine environment. They may be able to survive in seawater for hours, months, or even years, although with reduced propagation rates. *Bacillus thuringiensis*, a bacterial subspecies widely employed on land as a biopesticide, is an example of a microorganism commonly transferred to the ocean by terrestrial runoff (Walton and Mulla, 1992; Meadows, 1993).

The second type, facultative marine bacteria, includes species whose metabolism is attuned equally well to terrestrial, freshwater, and marine environments. Certain species of *Pseudomonas* and *Vibrio* are examples of organisms able to thrive in diverse environments.

The third type, obligate marine bacteria, are comprised of bacteria that have an absolute requirement of sodium for growth and/or osmotic functions (MacLeod, 1965). Typically, in obligate marine bacteria, the transport of solutes into the cell is sodium dependent (Jenson and Fenical, 1994). For example, studies have demonstrated that there is an absolute requirement of sodium for growth of *V. cholerae* (Singleton, et al., 1982). In addition, toxin production of this vibrio is markedly enhanced by a high salinity when concentration of other nutrients is low (Tamplin and Colwell, 1986). Another interesting example of sodium dependent organisms is the recently identified obligate marine bacterium *Pseudoalteromonas* species F-420. This organism has been found to produce an antibiotic that inhibits marine gram-negative bacteria, but is inactive against tested terrestrial microorganisms (Yoshikawa et al., 1997).

For the purpose of risk assessment and management in the marine environment the issue whether the bacterium under consideration is of an obligate marine species or not is of little practical consequence. Rather, the issue is whether it is able to survive long enough in saltwater to directly affect ecosystem function or influence indigenous microorganisms through transfer of genetic material. If the bacterium under consideration has one or both of these capabilities, it is designated "marine" for the purpose of this study.

Marine bacteria have adopted a variety of survival strategies, including floating freely as bacterioplankton in the water column, living in symbiotic association with macroorganisms, existing as a component of biofilm on natural and manmade surfaces, and inhabiting marine sediment.

Seawater everywhere supports large numbers of free-floating bacterioplankton. However, seawater containing increased amounts of organic matter supports larger populations of heterotrophic marine bacteria. For example, coastal areas receiving significant domestic, industrial, and agricultural runoff contain larger numbers of heterotrophic bacteria than open ocean regions (see Plate 2, p.53).

A large number of marine bacterial species coexist with macroorganisms as symbionts; i.e., in commensal, mutual, or synergistic relationships. For example, bacteria growing in tissues of sea sponges release secondary metabolites into the surrounding water as a chemical defense that dissuades predators or competitors from invading the host's living space (Schmitz, 1994; Althoff et al., 1998). By protecting the host, the symbiotic bacteria secure their own habitat (Santavy, 1995).

Every object immersed in seawater quickly attracts microorganisms, which attach to its surfaces, forming a biofilm. The biofilm, in turn, provides a base upon which larger life forms can attach. Although biofilms tend to have similar structures and functions wherever encountered (Potera, 1996), only a few of the attaching marine bacterial species constituting biofilms have been isolated and characterized. Bacteria and eukaryotic organisms involved in biofouling have been the subject of intense research in the last decade.

Similar to terrestrial soil, marine sediments and coastal region mudflats are inhabited by a variety of both anaerobic and aerobic bacterial species. A substantial proportion of bacteria inhabiting marine sediments are anaerobic, methanogenic bacteria that reduce CO_2 to CH_4 (methane), in the process using H_2 as an electron donor. Methanogenic anaerobic bacteria found in marine sediments are, therefore, important consumers of CO_2, one of the greenhouse gases. As is the case with all marine microorganisms, many of the species found in marine sediments have yet to be characterized. Moreover, despite the similarities, the microenvironment of marine sediment may be quite different from that of terrestrial soil (Fenical, 1993).

As with terrestrial microorganisms, gene flow occurs among marine bacteria or, possibly, between bacteria and other types of marine microorganisms. The three principle mechanisms of gene transfer known to operate in the terrestrial environment—conjugation, transduction, and transformation—can be assumed to function similarly in the marine environment (Jackman et al., 1992; Miller and Sayler, 1992; Stewart, 1992).

Conjugation is the mechanism whereby two cells directly exchange genetic material. In certain environmental niches bacterial cells are located in close proximity to one another, thereby providing conditions conducive to conjugation. In the marine environment, particularly favorable conditions are found in biofilms, biomats, digestive tracts of macroorganisms, and digestive vacuoles of dinoflagellates. Seawater itself is a continuous environment in which marine microorganisms collide with one another, enabling them to conjugate.

The degree of efficiency of conjugation depends to a large extent on how closely the two cells that come in contact are related (Davis, 1980). If the two cells belong to the same species, e.g., both are *Escherichia coli* cells, then conjugation often works efficiently. It functions well, but less so, if the two cells belong to the same genus, e.g., *Enterobacter aerogenes* and *Enterobacter cloacae*. Exchange efficiency declines to a marked extent, although conjugation may still occur, between cells of different genera in the same family, e.g., *E. coli* and *Salmonella* of the family *Enterobacteriaceae*. Conjugation is unlikely to take place between cells of distantly related or unrelated families, e.g., bacteria and microalgae. Effective conjugation also depends on signalling; i.e., the set of complex chemical messages that one cell uses to communicate with another (Cotter and Miller, 1996). Very little is known about this phenomenon, but it is reasonable to assume that the frequency and quality of conjugation depends on signals exchanged by the bacterial cells.

Transduction is exchange of genetic material between two cells through a vector, typically a phage. It is known that transduction occurs with some frequency in natural soil (Levy and Marshall, 1998; Meadows, 1993). In view of the large number of free viruses in the marine environment (see below), it appears that bacteria would have many opportunities to exchange genetic material by transduction, and this possibility must be considered in any assessment of ecological risk (Proctor and Fuhrman, 1990). In 1997, a study for the first time demonstrated genetic transfer via transduction between marine bacteria and enteric bacteria (Chiura, 1997). This study also indicated that viruses may have an important role in controlling bacterial populations in the oceans.

Transformation is the process whereby a cell takes up genetic material, such as plasmids or naked DNA, directly from the immediate substrate. For this to occur, both free DNA and competent cells (i.e., cells capable of taking up DNA) must be present. In addition, the internal conditions of the competent cells must be such as to allow the DNA to become stably integrated in the genome (Pickup, 1992). Competence appears to be a regulated state in transformable bacteria; the organisms stimulate the expression of phenotypic characteristics required for competence induction (Stewart, 1992). Research has demonstrated that soil bacteria, e.g., *Bacillus subtilis*, commonly utilize transformation. However, little is known about the frequency or importance of transformation among bacteria in the marine environment. Nevertheless, free DNA is known to exist in significant quantities in seawater, although it can be rapidly destroyed by DNA-hydrolyzing bacteria (Pickup, 1992; Stewart, 1992;

Lorenz and Wackernagel, 1994; Jeffrey et al., 1990). Because high molecular weight DNA and many transformable bacteria are present, some authorities believe transformation is a relatively common occurrence in the marine environment (Lorenz and Wackernagel, 1993).

Microcosm and mesocosm studies provide evidence that gene transfer between introduced transgenic bacteria and wild types is not an uncommon occurrence in seawater and marine sediment. It has been found that the frequency of gene transfer by any of the three mechanisms described above is dependent on biotic characteristics. These include the concentration of competent cells (overall concentration in seawater is circa 10^6 to 10^8 cells per ml; however, the concentration of competent cells is not known); phage concentration (circa 10^7 to 10^9 viruses per ml); and the concentration of free-floating DNA (from a low of 0.2 mg/liter in offshore water to a high of 44 mg/liter in estuarine water). Physical and chemical factors also play a role. Physical forces bring cells, phages, and DNA particles into contact, and chemical factors, such as salinity, pH, temperature, nutrient concentrations, etc., contribute to the effectiveness of gene transfer (Lorenz and Wackernagel, 1994, 1993). Exactly how these factors interact to affect the probability and frequency of the exchange of genetic material is not known.

Methods used to investigate DNA (and RNA) in environmental samples include plasmid analysis, nucleic acid hybridization, and polymerase chain reaction (PCR) (NRC CMMB, 1994; Pickup, 1992; Sayler et al., 1992). The first method, plasmid analysis, involves examining cells for plasmids containing marker genes, e.g., those coding for antibiotic resistance or light production. The second, nucleic acid hybridization, employs DNA probes to detect sections of DNA sequences in a sample that are specific for a given bacterial species. The third, PCR, involves amplifying the nucleic acids present in samples to obtain sufficient quantities to identify the origin of recovered DNA sequences. By applying these methods, it is possible to determine whether gene transfer occurs between different species in the marine environment, the extent of gene transfer, and the origins of a gene, DNA sequence, or RNA sequence.

This section on marine bacteria would be incomplete if the phenomenon of "viable but nonculturable" (VBNC) bacteria was overlooked. In the early 1980s, the existence of this previously unrecognized class of bacteria was confirmed (Xu et al., 1982, 1984; Colwell et al., 1985). The fact that as many as 200 to 5,000 times more bacteria could be seen through the microscope in a given quantity of seawater than could be cultured had been known for many years (Zobell, 1946). However, whether the nonculturable bacteria were viable was in dispute until newer tests could be applied (Huq and Colwell, 1995). It is now known that the majority of nonculturable marine bacteria exist in a VBNC state; i.e., the organisms are viable but are in a dormant state and cannot be cultured employing standard microbiological technology. Innovative methods have been developed over the past ten years that allow for accurate assessments of viability (Colwell and Huq, 1994). Although it is not yet clear why bacteria enter the VBNC state, it has been determined that the VBNC phenomenon is under genetic control (Oliver, 1993; Hill et al., 1996). Efforts to overcome barriers to characterizing all bacteria in seawater are important since it has been determined that some species pathogenic to humans, including *Vibrio cholerae*, are able to enter the VBNC state (Hugenholtz and Pace, 1996).

2.3 Cyanobacteria

Cyanobacteria, the Cyanophyta, are oxygenic, photosynthetic microorganisms. Cyanobacteria exhibit coloration due to phycobiliproteins, with their bluish coloration imparted by the pigment phycocyanin. The cyanobacterial cell wall, ribosomes, and nucleic acid structure are similar to those of bacteria. Although some scientists have surmised that cyanobacteria are intermediate between prokaryotic photosynthetic bacteria and eukaryotic microalgae (Becker, 1994), others disagree, noting that the cyanobacterial pigments and genomics do not support that hypothesis (Schneegurt, 1997).

Cyanobacteria are ubiquitous, existing as either free-living forms or symbionts in soil, freshwater, and seawater. Some of them are adapted to survive in extreme environments; for example, cyanobacteria populate both alkaline hot springs and Antarctic rocks (Skujins, 1992). Collectively, cyanobacteria are significant primary producers and the most important food source for many marine micro- and macroorganisms, from zooplankton to finfish. Some cyanobacteria, in particular members of the genus *Tichodesmium*, fix nitrogen and therefore affect the biological productivity of large expanses of the ocean (Capone, et al., 1997).

Under certain conditions some cyanobacterial species are able to bloom, i.e., reproduce rapidly, increasing in number to several million per ml of seawater (Carmichael, 1994). Blooms of cyanobacteria (and other marine microorganisms) occur frequently, especially in warm water. Most blooms are innocuous. At times, however, a toxin-producing cyanobacterial species may bloom. When this occurs, nearby aquaculture facilities or fisheries may suffer heavy damage; the toxins produced by the cyanobacteria may kill fish and shellfish directly or make harvested animals unfit for human consumption because of accumulated toxins in the tissue. The most common toxin-producing cyanobacterial genera are *Microcystis*, *Oscillatoria*, *Anabaena*, and *Aphanizomenon* (Carmichael, 1994; Mitchell, 1996).

Cyanobacterial masses also may form mats that float on the ocean or stick to the surfaces of materials in the water. Since these mats are difficult to disperse and tend to attract other adhering organisms, they create serious fouling problems in harbors, aquaculture facilities, and on equipment operating in the sea. Microbial mats, however, may also be beneficial. For example, mat-forming organisms from the genus *Spirulina* have been cultivated since ancient times in Mexico as food. Today, *Spirulina* is grown in much the same way as microalgae to produce health food and other high-value products including vitamins, nutritional supplements, and industrial pigments (Becker, 1994).

2.4 Fungi

Both marine and terrestrial fungi can be recovered from seawater and ocean sediment. Marine fungi complete their life cycles in the oceans, but seawater often inhibits spore germination of terrestrial fungi. Marine fungi, for convenience, can be subdivided into three classes: aureniculous species (which grow in beach sand), lignicolous species (which grow on submerged wood), and caulicolous species (which inhabit decaying marine plants). Marine fungi have various characteristics that may be useful or harmful. For example, several marine fungal species, primarily aureniculous species, are able to utilize hydrocarbon compounds such as petroleum as their main carbon source and, therefore, may have utility as bioremediators (Cooney et al., 1993). Conversely, some marine fungal species cause serious diseases in animals and plants (Fenical and Jensen, 1993).

Marine fungi, in general, are poorly characterized, partly because they have not received much attention and partly because they are difficult to isolate and culture. By 1991, only 321 marine fungal species had been partially or completely characterized, while over 69,000 terrestrial fungi had been described (Bernan, 1995). Although it is not known how many fungal species exist in nature, there probably are as many, if not more, in the marine environment than in the terrestrial environment (Groombridge,

1992). Research involving marine fungi is likely to increase sharply in the next few years because more and more investigators are becoming aware of the ability of these organisms to produce interesting metabolites in culture.

2.5 Meiofauna

Meiofauna is a collective term for microscopic invertebrates that inhabit beaches, marshes, mudflats, and other coastal areas washed by seawater (Squires, 1997). The most common meiofauna are nematodes and gnathostomulida. Other organisms that fall into this category are ciliates, copepods, polychaetes, and rotifers. Collectively, these organisms are thought to have an important role in breaking down organic matter, thus helping to keep shore areas clean. Meiofauna also are an important part of the food chain. After they are swept out into the sea by waves, they become food for juvenile fish, shellfish, worms, and other marine macroorganisms. Very little is known about the relationships between meiofauna and the other types of marine microorganisms described in this chapter.

2.6 Microalgae

Algae constitute a highly diverse group of aqueous plants ranging in size from the microscopic, unicellular plankton, to giant kelp 50 meters long (Meints and van Etten, 1992; Tver, 1979). Only microalgae, which resemble marine bacteria in size and shape, are considered here.

Microalgal species inhabit every ocean of the planet, as well as land, lakes, and rivers. Microalgae vary greatly in color and shape from species to species. Major types include diatoms (*Bacillariophyceae*), which possess siliceous shells, and haptophytes, which are covered with calcified scales.

The importance of microalgae to all life on earth cannot be overstated. Collectively they are the major primary producers of the ocean; i.e., they convert atmospheric CO_2 into organic substances that are food for heterotrophic microorganisms, invertebrates, and fish (Fenical and Jensen, 1993). Furthermore, microalgae collectively affect atmospheric and meteorological conditions. For example, they generate oxygen, produce chemical substances that promote cloud formation, and control the ocean's carbon cycle (Staley et al., 1997; Watson and Liss, 1998; Azam, 1998).

Approximately 1.3% of all microalgal species produce toxins. Similar to cyanobacteria, some microalgal species are capable of blooming. If it was to happen that a toxin-producing microalgal species blooms, and if that event was to take place in the coastal zone, people might get hurt and industry damaged. Directly, the blooming microalgae can cause massive fishkills of both wild and aquacultured fish. It also is possible that toxins accumulate in the tissue of shellfish inhabiting the affected areas. If contaminated shellfish are harvested and eaten, consumers are likely to contract amnesic shellfish poisoning. Another possibility is that toxin-producing microalgae are eaten by zooplankton, which in turn are eaten by fish fry, and so on. At each stage in the foodchain, the toxins produced by the original microalgae become more concentrated, eventually reaching a high concentration in the tissues of the predators constituting the top of the food chain, such as barracuda and yellow tail tuna. Contaminated fish present a public health hazard since they may cause paralytic, diarrhetic, neurotoxic, and histamine and ciguatera fish poisonings among human consumers (Epstein et al., 1994; Leffler, 1998).

The number of microalgal blooms recorded annually throughout the world is increasing (Tester, 1994; Broad, 1996; Leffler, 1998). Although the reasons for this rise are not known, three sets of factors might be responsible. First, wetland loss, erosion, over-application of fertilizers, and other unsound agricultural practices have caused increased eutrophication of the coastal zone (Patz et al., 1996). Second, global warming has elevated average sea surface temperatures approximately 0.7° Celsius over the last hundred years. The warmer temperatures appear to create more favorable conditions for blooms (Patz et al., 1996), possibly by affecting the balance between phytoplankton and their predators, the zooplankton, in favor of the former. Third, newly emergent species or strains of microalgae may be responsible for some of the blooms (Wood et al., 1993).

2.7 Plankton

Plankton is an umbrella term that encompasses all types of microscopic organisms that travel with water movement or swim slowly in the water column. As such, plankton constitute the bulk of marine biomass. Furthermore, nearly all life forms in the oceans spend part of their life cycle as plankton (Wiebe et al., 1996). (Microscopic forms constituting parts of the life cycles of macroorganisms, including fish and shellfish, are considered in greater detail in Chapter 3).

Five types of plankton predominate: (1) phytoplankton—microscopic plants, including diatoms, haptophytes, and other microalgae; (2) zooplankton—microscopic animals, including protozoa, worms, and larvae of benthic and nektonic animals; (3) bacterioplankton—marine bacteria; (4)

ichthyoplankton—microscopic fish larvae; and (5) hydroids—the polyp stage of hydromedusan jellyfish. Usually a substantial proportion of plankton is made up of phytoplankton—primary producers occupying the bottom of the food chain. Zooplankton and ichthyoplankton graze on phytoplankton and, in turn, are predated upon by hydroids and clupeoid fish (e.g., anchovies, herring, menhadens, sardines, sprats, etc.) (Jeffrey et al., 1994).

2.8 Protozoa

Protozoa are microscopic, one-celled animals possessing organelles, but no special organs for digestion, circulation, or reproduction. They may exist either as sediment dwellers, free-swimming plankton, or as parasites attached to, or living within, larger host organisms. Of potential importance to risk assessors is the finding that the digestive vacuole of protozoa offers an ecological microniche conducive to gene transfer between bacteria ingested by the host organism (Schlimme et al., 1997).

Important members of the protozoa phylum are dinoflagellates, characterized by possession of two flagella. These organisms constitute a substantial proportion of the zooplankton populations of the world's oceans. Some dinoflagellates are bioluminescent, and a few dinoflagellate species are capable of blooming (Glasgow et al., 1995). Of the dinoflagellates known to bloom, the toxin-producing species *Gymnodinium breve* is the most infamous. This organism, which carries red photosynthetic pigments, causes the surface of the water to turn red (thereby creating a "red tide"). A bloom of the *G. breve* strain may also produce fishkills, either through direct contact with toxins or through oxygen depletion in the water. Another dinoflagellate known for its adverse effects is *Pfisteria piscicida*, an organism that secretes a toxin into the surrounding water for the purpose of stunning fish so it can feed on them (Burkholder and Glasgow, 1997). In 1997, *Pfisteria* was identified as the cause of a fishkill of approximately one million menhaden in North Carolina; other massive fishkills due to this organism have been observed in Maryland (Greer, 1997).

2.9 Viruses

Viruses are parasites that grow and multiply only within host cells (Stent, 1963). Research employing transmission electron microscopy and epifluorescence microscopy has demonstrated the presence of a large number of submicron particles in the ocean surface layer, including free viruses, defective phages, phage ghosts, and colloid assemblies (Proctor and Fuhrman, 1990; Suttle et al., 1990; Wommack et al., 1992; Epstein et al., 1994; Proctor, 1997). Of these, the most important are free viruses, comprising bacteriophages, archael viruses, and eukaryotic viruses (Proctor, 1997). Viral abundance in surface seawater varies widely, from 10^6 to 10^{11} per liter (Proctor and Fuhrman, 1990; Fuhrman et al., 1993). The role of viruses in surface waters is unclear, but several investigators have proposed that most are phages, i.e., viruses that infect bacteria, plankton, and microalgae (Miller and Sayler, 1992). If most marine viruses are phages, they probably have an important role in limiting the populations of microorganisms in the oceans, thereby regulating productivity of marine micro-ecosystems (Suttle et al., 1990; Proctor, 1997). The finding that as many as 70% of heterotrophic marine bacteria are infected by viruses, causing between 10 and 100% of bacterial mortality (Proctor and Fuhrman, 1990), supports this hypothesis.

Although viruses usually are perceived as pathogens, some may be beneficial to their hosts. For example, phages whose genomes become integrated with the genome of the bacterium *Bacillus subtilis* have been found to contribute genes that assist the host cell in resisting the toxic effects of harmful chemicals, including heavy metals (Williams, 1997). It may be that other marine viruses confer an as yet unknown advantage on their hosts.

Viruses pathogenic for macroorganisms may persist outside the host cell in seawater as free viruses. The sensitivity of viral strains to chemical and physical stresses varies widely. For example, *Baculovirus penaei*, a shrimp pathogen, can survive for seven days in 22°C seawater and 14 days at 5°C; while an enterovirus survives for up to 90 days at 3 to 5°C but, at most, 9 days at 25°C (Hegde et al., 1996).

Filter-feeding shellfish may accumulate viruses, including human pathogens such as the oyster-borne Hepatitis A virus and calciviruses (Norwalk-like viruses) causing gastroenteritis (Anonymous, 1994; Smith et al., 1998). Since viral pathogens carried by shellfish commonly cause gastrointestinal and hepatic infections, public health officials of coastal states monitor fisheries continuously for contaminated shellfish.

3. POTENTIAL GEMM APPLICATIONS

Theoretically, any of the marine microorganisms described above may be subjected to genetic engineering. However, for the sake of practicality, only GEMM strains that are likely to be developed for use in the open marine environment in the next five years, 1999 to 2004, are considered in detail here. Accordingly, this section has two parts. First, the categories of marine microorganisms described above are reviewed and their potential as subjects for genetic engineering are briefly assessed. Second, those organisms likely to be developed into useful products within five years are described and analyzed more fully.

3.1 Marine Microorganisms as Research Subjects

As part of an earlier survey, data was gathered on the types of marine microorganisms being employed in research with the intent of developing GEMM applications. The results were entered into a database dedicated to marine biotechnology called MARBIO (Zilinskas et al., 1995). A summary of relevant survey results pertaining to each type of marine microorganism, augmented by more recently published information, follows.

Archaea. Much research in marine biotechnology is being done to obtain natural products produced by archaea, especially enzymes functioning at high temperatures (Adams, 1995). Attempts are being made to engineer archaea to produce larger quantities of desirable products in the laboratory. However, for the present there are no indications that archaea species are being developed for uncontained uses.

Bacteria. Many research projects are being performed throughout the U.S. to develop various types of genetically engineered marine bacteria (GEMB), usually for bioremediation and other environmental applications. Therefore, a more extensive discussion of the possible uncontained uses of genetically engineered marine bacteria is provided below.

Cyanobacteria. The first transformable cyanobacterium, *Anacystis nidulans* R2, was developed in The Netherlands in the late 1960s and early 1970s (Schneegurt, 1997). In the U.S. the genetic engineering of the cyanobacteria *Synechocystis* 6803 was reported first in 1983 (Williams and Szalay, 1983). Information from the MARBIO survey and recent literature indicates that most of the research in genetic engineering employing cyanobacteria is being done as part of basic research, such as determining the nitrogen fixation (nif) gene sequence and the photosystem sequences (Zilinskas et al., 1995; Skujins, 1992). In addition, some applied research is being done to develop cyanobacteria to overproduce pigments and chemicals useful for medicine and special purposes, e.g., to treat waste water or to bioremediate soil (Ciferri et al., 1989; Elhai, 1994; Mhiri and De Marsac, 1997). However, progress to develop these organisms for environmental applications has, in general, been slower than with other bacterial systems. With the exception of one possible application of transgenic cyanobacteria as a pesticide (see below), research on cyanobacteria is unlikely to lead to uncontained uses within the next five years.

Fungi. Marine fungi with biodegrading capabilities have been suggested for bioremediation applications (Cooney et al., 1993). However, no transgenic marine fungal species has been developed for uncontained use and such a development is unlikely within five years.

Meiofauna. None of the microscopic invertebrates that constitute meiofauna are being developed for uncontained applications.

Microalgae. Native marine microalgae, including *Dunaliella*, *Chlorella*, and *Scenedesmus*, are being used to manufacture speciality chemicals destined mainly for the health food or speciality chemical industries. Although transgenic forms of microalgae are being developed for more efficient processing of these products (Becker, 1994; Radmer, 1996), there are no indications that transgenic microalgae will be used in uncontained applications in the next five years.

Protozoa. While the insertion of genes coding for the production of *Bacillus thuringiensis* toxin into the freshwater protozoa *Tetrahymena pyriformis* has been reported (Gelernter and Schwab, 1993), there is no information indicating that marine protozoa are being developed for uncontained applications.

Viruses. In the terrestrial environment baculoviruses have been developed for use as biopesticides (NRC, 1989; UKDOE, 1995). These may be transported into the oceans through runoff but are unlikely to pose ecological risk (Cory and Hails, 1997). Marine viruses are being developed as vaccines for aquaculture (Leong et al., 1997; Benmansour and de Kinkelin, 1997), but these applications are contained. Otherwise, there is no information to indicate that marine viruses are being developed for uncontained applications.

In summary, no genetically engineered archaea, fungi, meiofauna, microalgae, protozoa, or viruses are likely to be used in an uncontained application in the marine environment within five years. On the other hand, strains of GEMB and transgenic cyanobacteria are under development, and soon some of these could soon be candidates for field testing.

3.2 Potential Applications for Transgenic Marine Microorganisms

Analysis of MARBIO survey results and information from recent publications leads to me to conclude that GEMB strains may be utilized in two types of environmental application: as biopesticides and for bioremediation.

3.2.1 Bacterial pesticides

The bacterial species *Bacillus thuringiensis* (Bt) has been employed as a biopesticide on land for more than ten years (Prieto-Samsünov et al., 1997; Tabashnik, 1994). In this application, a solution containing Bt is sprayed over the plants to be protected. Larvae of butterflies, moths, mosquitoes, black flies, and other insects feed on leaves contaminated with Bt, in the process ingesting the bacteria. Bt produces toxins that are activated by the alkaline solution present in the gut of the larvae. The toxins dissolve the gut lining, killing the insect pests.

To enhance the range and effectiveness of Bt toxins, researchers have inserted genes coding for Bt toxin production into aquatic cyanobacteria and protozoa (Gelernter and Schwab, 1993). For example, in 1995 the Cyanotech Corporation in Hawaii employed genetic engineering to remove the toxin gene from Bt and insert it into the cyanobacterium *Synechococcus* (Anonymous, 1995). Since mosquito larvae feed on cyanobacteria in nature, and since cyanobacteria are normal, ubiquitous inhabitants in freshwater environments, the newly developed transgenic cyanobacteria may serve as an environmentally safe biopesticide that can replace toxic organic chemical pesticides now used for mosquito control. While this application is designed for freshwater use only (mosquito larvae grow only in freshwater), rivers or streams into coastal waters could transport the newly developed transgenic cyanobacteria. If so, they would, for practical purposes, become GEMMs released into the open marine environment, and risk assessors would have to consider any associated ecological or human health risks.

3.2.2 Bioremediation

Microorganisms will attack almost any complex chemical substance released by natural or human action into the terrestrial or aquatic environments. Many microbes are able to secrete exoenzymes that break down complex molecules. This process is called biodegradation. Biodegradation often is a slow process; for example, the breakdown of raw petroleum spilled from an oil tanker onto a beach may take several years or, in cold climates, decades.

Chemicals may also be removed from the environment by bioaccumulation; i.e., some microorganisms are able to absorb chemicals through the cell wall and accumulate them within special intracellular structures. For example, some bacterial strains absorb and accumulate metals, including heavy metals, which usually are toxic to other life forms. The sequestered toxic metals thus are removed from the environment where they might otherwise cause harm.

In contradistinction to biodegradation and bioaccumulation, bioremediation is the deliberate application of biological activity to reduce the concentration of toxicity of contaminants (US Congress, 1991; Atlas, 1995). Bioremediation applies technology to enhance natural rates of biodegradation or bioaccumulation of pollutants in the soil, on shorelines, floating on the surface, or embedded in marine sediment. Industry has utilized bioremediation to treat contaminated sites at least since the early 1940s when the American Petroleum Institute began tracking these activities (API, 1995). The most common approach to bioremediation is to enrich the polluted site with fertilizer to encourage growth of biodegrading and bioaccumulating microorganisms that are naturally present (US Congress, 1991).

Under natural conditions, the number of native microorganisms present at a given locale is limited by the amount of available feedstock and cofactors—nutrients such as nitrogen and trace elements vital for metabolism. While most manufactured chemicals discharged during a pollution event are likely to be amenable to biodegradation or bioaccumulation, the discharge usually will not contain cofactors. Thus, an overabundance of feedstock suddenly appears at a site, but the amount of cofactors available remains constant. Without an increased supply of cofactors, the ability of indigenous biodegrading/bioaccumulating microorganisms to multiply would remain as before, and the rate of biodegradation/bioaccumulation would remain low. To overcome this barrier, pollution control engineers scatter fertilizer over the site slated for bioremediation. Typically, the fertilizer is supplied by a bioformulation company that has developed proprietary blends of nutrients and trace elements designed for specific situations (Sybron Biochemical, no date). Judicious fertilization of approximately 120 kilometers of Alaskan coast line fouled by petroleum spilled from the Exxon *Valdez* facilitated more complete and more rapid cleaning than would have occurred if nature had been allowed to take its course (Bragg et al., 1992; Prince, 1997).

A second approach is to seed the polluted area with biodegrading or bioaccumulating microorganisms. Dozens of firms in Europe and the U.S. develop and sell bioremediation products, usually blends of naturally occurring biodegrading or bioaccumulating microorganisms whose parent strains originally were isolated from near and distant sites polluted by various substances (Cutter Information Corporation, 1997). Parent strains typically are used as seed cultures for mass production of microorganisms constituting the commercial blends. Firms sell blends designed for specific applications; for example, blend A might be touted as being effective in treating saw mill wastes, blend B for cleaning ship bilges, and blend C for cleaning up sites contaminated with organophosphate pesticides. Some blends may have been developed for even more specialized applications; for example, blends are marketed by developers who assert that they are effective only within narrow temperatures ranges.

The quantities of naturally occurring microorganisms being prepared for uncontained environmental applications are huge. For example, since the mid 1980s, Sybron Chemicals Inc. in New Jersey, which may be the largest supplier of microbial cultures for environmental applications in the U.S., produces and sells an average of one million pounds of dried microbial biomass and two million gallons of microbes suspended in liquid per year (Sybron Biochemical, no date). The cell count of the dried product is in the range of 10^8 cells per gram. Although *Bacillus*, *Enterobacter*, and *Pseudomonas* strains are the chief ingredients, up to 30 different bacterial strains originally recovered from a variety of sites worldwide may be used in these products.

The effectiveness of seeding with bioremediating microorganisms, as opposed to applying fertilizer to encourage growth of naturally occurring organisms, is disputed, however. Field tests and analyses of commercial applications by independent researchers have yielded mixed results. Sites along the Gulf of Mexico in Texas contaminated by oil spilled from the *Mega Borg* in 1990 were treated with microorganisms, but benefits could not be unequivocally attributed to these efforts (TGLO, 1992). Findings from field experiments involving polluted coast lines in Prince William Sound and Delaware Bay indicated no benefit from seeding (Venosa et al., 1992, 1996, 1997; Prince, 1997). However, limited success has been claimed for attempts to bioremediate contaminated sediment in a canal at Zierikzee, The Netherlands (De Meyer et al., 1997), and at Eagle Harbor in Puget Sound (Anonymous, 1996c).

The lack of consensus on whether bioaugmentation is effective may be due to the significant technical problems that may attend bioremediation employing biodegrading or bioaccumulating bacteria. It could be that in one case a bioremediation project encounters no problems, so its results are positive. Conversely, another project may be beset by difficulties, so it fares badly. In practice, five major problem areas have been identified:

(1) A pollution event usually involves several compounds with differing chemical structures. However, biodegrading and bioaccumulating bacteria tend to be specific as to their action; therefore, no single bacterial species is able to bioremediate more than one or a few compounds at a polluted site.

(2) Biodegrading bacteria utilize toxic compounds as carbon and energy sources. If the concentration of these compounds is too low, biodegraders cannot produce the enzymes that break them down. If too high, the pollutants may adversely affect the biodegrading bacteria.

(3) A pollution event may contain chemicals that are incompatible with the biodegrading or bioaccumulation process or may adversely affect the biodegraders and bioaccumulators.

(4) Some compounds may be adsorbed rapidly into soil, sediment, and/or water, thereby becoming too diluted to drive biodegradation.

(5) If any of the foregoing problems are present, the rate of bioremediation of pollutants at a site may be too slow to provide useful cleanup.

Molecular biologists may apply genetic engineering to overcome these problems. Six approaches bears noting. First, work is being done to develop cloning vectors that are specific for environmental applications; i.e., vectors that are stably maintained, nontransmissable, and environmentally benign (Timmis et al., 1994). Second, since biodegradation usually involves many enzymatic reactions within the bacterial cell, some of which may be rate limiting, investigators are identifying enzyme reactions that constrain performance and designing mechanisms to overcome these limits. For example, genes that control enzyme reactions may be made more stable and given improved kinetic properties (Timmis et al., 1994). Third, some enzymes involved in biodegradation are unstable or have suboptimal structures. Also, most enzymes is highly specific. Thus, genetic engineering can be used to construct genes coding for the production of novel enzymes or enzymes that have multifunctional activity. Host bacteria that receive these genes would be able to break down a greater range of chemicals than their wild relatives (Timmis et al., 1994). Fourth, scientists can manipulate genes for the purpose of assembling new or improved metabolic routes, or pathways, for breaking down compounds (Timmis et al., 1994; Mhiri and De Marsac, 1997). Fifth, researchers can genetically engineer bacteria to withstand environmental stresses that characterize particular sites (e.g., extremes in temperature or pH, low or high oxygen tension, inadequate availability of cofactors, etc.) (Timmis et al., 1994). Sixth, bacterial strains may be genetically engineered to express novel metal transport systems. For example, recently an *E. coli* strain received an insert from the yeast *Saccharomyces cerevisiae* that enabled the host cell to accumulate larger quantities of mercury than its wild counterpart (Chen and Wilson, 1997).

No detrimental effects on the environmental or human health have attended bioremediation carried out through fertilizing or seeding with indigenous bacteria (US Congress, 1993). The reason is that fertilizers, being made up of nutrients already present in the ocean environment, quickly become so diluted that they cannot cause harm. Indigenous microorganisms, even if applied in large quantities, decrease rapidly in number once the pollutant being treated is exhausted. At that time, their numbers revert to natural levels.

It is more difficult, however, to predict effects that may result from applications of genetically engineered microorganisms, be it on land or in the sea. There are many scientific unknowns and little useful prior experience with uncontained GEMM applications for bioremediation. It was not until March 1996 that the EPA gave permission for a genetically engineered *Pseudomonas putida* (strain IPL5 with a genetic insert from *Alcaligenes eutrophus*) to be used to bioremediate a terrestrial site contaminated by polychlorinated biphenyls that leaked from four large electrical capacitor banks (Sayler, 1995; Sayler and Sayre, 1996). This application, although it is being conducted in the open air, is contained by physical barriers. As this is written, no interim results have been published from this test. In any case, we cannot assume that findings from a terrestrial field trial are applicable to the marine environment.

4. RISKS ASSESSMENT OF UNCONTAINED GEMB INTRODUCTIONS

I make certain assumptions about steps that a developer will take before approaching a regulatory agency for permission to conduct an uncontained application of a newly developed transgenic marine bacterial species and the information the agency would require before giving its approval. Surmising that the most likely first application of GEMB will be for bioremediation, I assume that a developer would have commenced research with the aim of enhancing the biodegrading capabilities of a naturally occurring biodegrading marine bacterial species, taking one or more of the approaches described above. However it was done, once laboratory studies indicate that the newly developed GEMB had a significantly enhanced capacity for biodegradation, and therefore held promise for commercialization, the developer would perform further studies in the laboratory to determine whether the GEMB evidenced pathogenicity or could cause environmental damage. If laboratory studies indicated that the GEMB strain was safe, the developer would wish to test it in the open environment. The developer would, at this time, approach federal and state authorities for permission to conduct the field test and would present data to support the application. What data would be necessary to secure a permission to proceed with the field test?

Since the prospective field test will involve releasing a genetically engineered bacterial strain into the environment, and since it is intended for bioremediation, the developer can be expected to approach the EPA for permission to conduct the field test. Furthermore, authorities from the state where the test is to be conducted undoubtedly will also have to give their consent. However, as discussed in Chapter 5, the regulatory situation is not clear in regards to field tests in the marine environment. For this reason, I do not address the USEPA's Points to Consider in the hypothetical case. Rather, I analyze the data requirements in terms of the three general issues that underlie all sets of Points to Consider developed by regulatory agencies: (1) familiarity with the organism to be tested and the environment into which it is to be introduced, (2) possibility of controlling the test organism after release, and (3) potential detrimental effects that the test organism might have on humans or the environment.

4.1 Familiarity with the Test Microorganism and Environment

To evaluate a proposal to introduce genetically engineered microorganisms into the marine environment, it is first necessary to consider the level of familiarity with the test organism, as well as the environment into which it is to be introduced (NRC, 1989). Three questions are asked: Is the organism similar to those used in past introductions? Is the target environment similar to those used in past introductions? Is the intended function similar to those used in past introductions? If the answer to any of these questions is no, then the organism under consideration should be regarded as not familiar.

Some terrestrial GEMs are so well characterized that they can be defined as familiar. For example, a transformed variety of *Rhizobium meliloti* is familiar because the naturally occurring form has a history of safe use over a period of 100 years and because the genetically engineered strains of this species have demonstrated environmental safety in small-scale field tests over a nine-year period (McClung, 1997). On the other hand, even well-described GEMs, such as the first bacterial species patented in the U.S. (Chakrabarty, 1981) or the strain of *Pseudomonas putida* being tested for terrestrial bioremediation by the Sayre group, do not meet the familiarity criterion because insufficient data are available on these organisms to enable a risk assessor to predict their effects or fate in the open environment. It follows, therefore, that since no strain of marine bacteria is as well characterized and well known as any of the strains noted above (with possible exception of the frank pathogen *V. cholerae*), no strain of GEMB would be able to satisfy the familiarity criterion at present.

As discussed in Chapter 2, little is known about marine microbial ecology. In this chapter, it is made clear that the pathways for gene flow in saltwater also are mostly unknown. When we add the uncertainties described in Chapter 2 to those noted in this chapter, it becomes apparent that the second criterion cannot be met; i.e., there is a lack of familiarity with the environment into which the proposed introduction would take place.

In view of the almost certain inability of the developer to meet the familiarity criteria pertaining to both test microorganisms and the environment into which they are to be introduced, as well as doubts about the similarity of the proposed application to past applications, the second issue must be considered: Can the candidate GEMB be adequately controlled or contained after introduction?

4.2   Controlling Released Microorganisms

On land, it is usually possible to ensure that introduced organisms do not escape from the test site. If considered superficially, the bioremediation of an area of ocean or shoreline might appear similar to the bioremediation of a terrestrial site. However, most terrestrial sites can be contained; something that is not possible with marine or shoreline sites. For this reason, risk assessors must assume that if a number of GEMMs are introduced onto marine or shoreline sites, some will escape. In other words, no physical methods exist for ensuring that GEMMs introduced into an area of polluted ocean, spread over a section of intertidal zone, or injected into sediments in harbors or estuaries will be restricted to the point of application. Instead, introduced GEMMs are likely to escape beyond the boundaries of the site onto which they were introduced and, thence, be transported horizontally by ocean currents, perhaps over long distances, or vertically through the water column to the sea floor. Once GEMMs have been introduced, it may be impossible to recapture or destroy all test organisms unless, perhaps, biological containment features are genetically engineered into the released organisms.

The concept of biological containment was developed in the early days of recombinant DNA research. The idea is that foreign genes should be introduced only into strains of bacteria that have been otherwise bioengineered for safety; i.e., bacteria that once transformed cannot transfer their genes to other bacteria or survive outside the laboratory environment (Schweder, 1996). Early safe strains were altered so that they required a particular chemical not available in nature for survival. Recent advances in biological containment include development of bacterial strains containing genetic constructs that cause the host cell to die under specified conditions. Such constructs typically include a gene that codes for production of a toxin lethal to the host cell and a promoter sequence that activates the toxin gene in response to a precise signal, such as a temperature change, or the presence or absence of a specific chemical or nutrient. For example, a recently developed suicide construct allows cells of a biodegrading strain of *Pseudomonas putida* to survive only in the presence of certain aromatic hydrocarbons it has been engineered to degrade (Szafranski et al., 1997).

There are two major problems with biological containment based on suicide systems. First, mutations may occur that deactivate the safety features. Second, no system is 100% effective in bacteria; some bacterial cells will survive even if the suicide system they contain remains intact and operates at high efficiency. Since billions of bacteria are introduced in any application, a very low survival rate will still leave significant numbers of introduced microorganisms extant, and since bacteria can reproduce rapidly, a population of survivors and offspring could quickly become established. In either case, the biological containment mechanisms would be rendered ineffective.

Scientists are developing novel strategies to improve suicide systems. The most promising approaches consist of inserting multiple self-destruction systems in each organism. For example, a group of researchers has developed a two-component suicide system; the first causes an essential vitamin to become completely depleted, while the second component produces lethal damage to cell membranes, walls, or nucleic acid. The survival rate of cells carrying this dual system is exceedingly low, between 10^{-7} and 10^{-8} of all cells per generation (Szafranski, et al., 1997). Some investigators believe this is still an unacceptably high survival rate. One authority has calculated that satisfactory biological containment cannot be achieved until introduced organisms carry eight separate suicide mechanisms, each with a different control factor (Atlas, 1998).

4.3 Potential Environmental and Health Effects of Introduced GEMMs

Since our knowledge of GEMMs and the marine environment is too scant to meet the familiarity criterion, and since containment of introduced marine microorganisms cannot be assured, potential adverse effects must be considered when assessing a proposed uncontained application. Our means for doing so, however, is limited since, as is seen below, past introductions of marine microorganisms, whether accidental or deliberate, have hardly been investigated. This being the case, I analyze whether useful lessons can be drawn from certain other past activities, including microcosm/mesocosm studies and past introductions of exotic organisms into the marine environment. I also discuss hypothetical adverse effects on human health.

4.3.1 Microcosm/mesocosm studies involving marine phenomena

In this section, I refer to five adverse effects that have attended introductions of exotic organisms in the terrestrial environment (Zilinskas and Lundin, 1993) and, using data from microcosm/mesocosm and other studies, analyze whether similar adverse effects may occur in the marine environment.

(1) Once an introduced microbial species colonizes a new locale, it may prove impossible to eliminate.

No long-term studies have been done on the persistence of introduced microorganisms in the marine environment. However, microcosms have been used in investigations of the survival of transgenic marine bacterial strains in the aquatic environment (Awong et al., 1990; Chao and Feng, 1990), survival of transgenic *Pseudomonas* in lake water (Morgan et al., 1992), survival of transgenic *Erwinia* in freshwater and their impact on other bacteria in the water column (Scanferlato et al., 1989), and degradation of various aromatics by transgenic *Pseudomonas* (Wagner-Döbler et al., 1992; Heuer et al., 1995) and *Alcaligenes* (Pipke et al., 1992). The general lesson that can be drawn from these microcosm studies, which involve a very limited number of transgenic marine bacterial species, is that the ability of transgenic forms to survive and persist in freshwater, saltwater, and marine sediments is at least equal to that of wild types (Scanferlato et al., 1989). These survival studies, however, are markedly affected by microcosm design, so results are not transferable from one microcosm design to another. Further, as noted above, the database for marine studies is limited.

(2) An introduced microorganism may disrupt local fauna through competition. In the worst case, introduction of a nonindigenous microbial species may lead to the extinction of one or more wild species.

I have found no studies that address this problem area. Undoubtedly, it would be difficult, if not impossible, to design a study that would isolate the disruptive potential of a single microbial species on local fauna. Even the best designed microcosm study can include only a few of the biotic and abiotic components of an environment that might be affected by an introduced strain of GEMM, leaving many more unexplored.

(3) Wild indigenous bacteria may suffer genetic degradation as a result of exchanging genes with the introduced species, or important genes may be lost if introduced species outcompete wild species, thus displacing native organisms at the site of introduction and throughout the dispersal range.

This area has been only partially addressed by microcosm/mesocosm studies. For example, the subject of gene transfer between marine microorganisms (Fulthorpe and Wyndham, 1991; O'Morchoe et al., 1988; Lebaron et al., 1993; Pickup, 1992; Pickup et al., 1993) and bacteriophage-bacteria interactions (Miller and Sayler, 1992; Wommack et al., 1998) has received substantial attention. Microcosm studies have employed *Vibrio* strain D19, *Vibrio* strain WJT-1C, and *Pseudomonas stutzeri* in such studies (Lorenz and Wackernagel, 1994). Saltwater microcosm studies have been reported in reference to gene transfers (Lebaron et al., 1993; Sörensen, 1993; Ashelford et al., 1997) and transfer of metal resistance (Gauthier et al., 1985). Studies have demonstrated transfer of plasmids between different *Vibrio* species. Microcosms have been constructed to study the characteristics of plasmids, and plasmid transfer, in marine sediment, air-water interface, and biofilm communities (Sandaa, 1993; Pickup et al., 1993; Lebaron et al., 1997; Dahlberg et al., 1997).

These studies demonstrate three points. First, introduced microorganisms, whether exotic or genetically engineered, are capable of exchanging genes with wild indigenous microorganisms. Second, no single organism can serve as a model for naturally transformable bacteria in a given habitat (Lorenz and Wackernagel, 1994). Third, plasmids isolated from different marine habitats have replication and/or incompatibility systems that are different from plasmids commonly used in plasmid biology (which are largely terrestrial) (Dahlberg et al., 1997). The issue of whether genetic degradation occurs directly as a result of gene exchange has not been addressed in microcosm studies, and it may be extremely difficult to set up an experiment to investigate this possibility.

(4) Introduced microbial species may be pathogenic to wild indigenous macroorganisms.

From studies in the terrestrial environment, we know that microbial pathogens possess common factors, such as the ability to produce toxins, adhesives, and invasins (Falkow, 1997). A variety of tests have been developed to detect factors possessed by bacteria indicative of pathogenicity, including agglutination of red blood cells and slime production (adhesins), assays to assess of penetration of cell monolayers (invasins), measurement of tobacco hypersensitivity (plant pathogens), and immunoassays to detect toxin antigens (USEPA and Agriculture and Agrifood Canada, 1997). Bacteria that are candidates for uncontained terrestrial applications usually are tested for pathogenicity in multihost pathogen systems, e.g., mice, plants, nematodes, amphibians, fish, and other life forms. Arrays of tests indicative of pathogenicity have been employed in preparation for all terrestrial field trials involving GEMs. As far as is known, no bacterial or viral species released to date on land has caused disease among organisms at or near test sites.

Analysis of the Ecological Risks Associated with Genetically 121
Engineered Marine Microorganisms

Some studies of diseases afflicting marine organisms have been carried out, usually as part of aquaculture practices. Thus, some bacterial species and viruses known to cause disease in marine animals and plants have been identified (Newman, 1993; Coelen, 1997), but most of these are important to aquaculture so their extent in the wild are not known. Of course, previously unknown marine pathogens are being discovered with some frequency. For example, a newly discovered strain of *Brucella* is the suspected cause of serious illness in seals, porpoises, dolphins, and otters (Jahans et al., 1997). Nevertheless, it is reasonable to assume that many other strains and species pathogenic to macroorganisms are as yet unrecognized. No microcosm/mesocosm studies have addressed the possible pathogenicity of introduced marine microorganisms.

The problems caused by the lack of knowledge about diseases in the wild are compounded by science having only a limited amount of information about the immune mechanisms of marine animals and plants. Again, most of this information comes from studies of animals important to aquaculture. It can be seen that we know very little about pathogens that attack wild marine life forms and the defense mechanisms of the host organisms.

In view of these gaps in knowledge about both marine microorganisms and macroorganisms, two observations may be made. First, experience with procedures for terrestrial microorganisms designed to determine pathogenic potential may not be applicable to marine microorganisms, and, second, the pathogenic potential of any GEMM proposed for uncontained application cannot be adequately tested at present.

(5) Introduced microbial species may disrupt local community structure, leading to ecological degradation.

From the terrestrial experience, we know that it is possible that microorganisms developed for bioremediation may injure wild organisms indirectly. Bioremediating organisms dispersed over a site will, by design, come into contact with the pollutant they have been developed to break down. A rapid increase in the population of introduced organisms is likely to ensue. In the process of breaking down the original pollutant, the introduced bacteria may produce intermediate metabolites or end products that are more toxic than the original substance. For example, bioremediation of trichloroethylene could produce the more toxic vinyl chloride, while degradation of the explosive TNT could generate the toxic residue toluene (USEPA and Agriculture and Agrifood Canada, 1997). It is also possible that the greatly increased population could produce more of the toxic metabolites or end products than the local environment can absorb. In this case, site monitoring may be unable to distinguish between the toxic effects caused by the original pollutant and toxic metabolites or end products resulting from bioremediation. Further, some bioremediating microorganisms are capable of producing exotoxins or endotoxins when stimulated by cues in the environment. For example, as noted above, certain vibrios require sodium before being able to produce toxin. When sodium is present, the effects of environmental toxins and toxic intermediates or end products produced by vibrios may be indistinguishable (USEPA and Agriculture and Agrifood Canada, 1997). Biologic toxins, like toxic chemicals, can directly kill or injure animals and plants in the area of the site undergoing bioremediation. It is also possible that toxins may be ingested by organisms low in the food chain, so when predators higher in the food chain feed on the contaminated organisms, they accumulate a higher concentration of toxins in their tissues. Predators on top of the food chain my be injured or killed if they eat animals containing high amounts of toxins

Several toxicity tests have been developed to determine safe concentrations of chemicals in soil or receiving waters, including the seven-day early life stage sheepshead minnow larval survival and growth test, the inland silverside larval survival and growth test, and the mysid survival growth and fecundity test (USEPA and Agriculture and Agrifood Canada, 1997). Other tests that might have applications in the marine environment are based on diatoms and bioluminescent marine bacteria, such as *Photobacterium phosphoreum*. The latter are used in the commercially available Microtox test, which signals the presence of toxic substances by a reduction in bioluminescence.

The toxic properties of a chemical can be readily detected and measured by standardized tests before release into the environment. However, if the agent being considered for release is a microorganism, its ability to cause disease to any animal or plant at the site of release often cannot be predetermined; existing tests for pathogenicity encompass only a few animal and plant species.

Unlike the terrestrial environment, the potential effects of introductions on marine microbial ecology have not been studied. As with the second problem areas described above, it would be extremely difficult to design studies to elucidate negative effects on complex ecosystems that exist in the open ocean, at the land-water interface, on shorelines, or in marine sediments.

4.3.2 Lessons from past introductions of non-native marine organisms

Enormous numbers of exotic marine macro- and microorganisms are continuously being introduced into sites throughout the world. The major source of exotic marine species introductions is international shipping. Ships carry organisms from one port to another in ballast water, bilge water, and hull encrustations. Much scientific attention has been focused on introduced exotic macroorganisms and their effects on the environment (Carlton, 1992, 1995; Lester, 1992; Committee on Ship's Ballast Operations, 1996). An example of a severely affected site is San Francisco Bay; it is estimated that macroorganisms of exotic lineage now outnumber indigenous forms (Stevens, 1996). Although the majority of introductions of exotic macroorganisms are innocuous (see Chapter 1), some introduced forms have caused enormous environmental and economic damage. The most notable recent example is the zebra mussel, *Dreissena polymorpha*, millions of which are clogging up industrial intake and outflow pipes in the Great Lakes region (Kastner, 1996; Johnson and Carlton, 1996).

In comparison, marine biologists have paid little attention to introduced exotic microorganisms. A few studies note in passing that zooplankton and diatoms are also part of the biological load carried by ballast water (Carlton, 1995; Geller, 1996), but as far as I have been able to discern, no long-term studies, with one exception, have been carried out to determine the fate or environmental effects of introduced microorganisms.

The one exception is the study of exotic viruses that cause shrimp diseases, including infectious hypodermal and hematopoietic necrosis syndrome, taura syndrome, white spot, and yellow head (JSA SVWG, 1997). Pathogenic viruses have been found in shrimp that supply U.S. aquaculture facilities, serve as bait, and are used as research subjects. Shrimp accidentally transported in ballast water also carries these viruses. These disease agents have caused epidemics in aquaculture facilities in Atlantic and Gulf of Mexico states, as well as among wild shrimp native to these regions. The newly introduced viral diseases have the potential to cause heavy damage to the shrimp processing and aquaculture industries in the U.S.

Just as the effects of accidentally introduced exotic marine microorganisms are generally unknown, the ecological impact of deliberately introduced microorganisms also has not been studied comprehensively. As noted above, blends of microorganisms produced by bioformulation companies have been used for bioremediation efforts, and these applications undoubtedly have resulted in the dispersal of large numbers of exotic bacteria over wide areas of U.S. coastal waters. Because their products enjoy proprietary protection, companies need not publish precise data on their blends. As a result, it is difficult for outside researchers to investigate their environmental effects, good or bad. In addition, applications of microbial blends are largely unregulated. The USEPA has no authority to control the use of naturally occurring microorganisms, unless they are used as pest control agents. In summary, little if anything has been learned from accidentally or intentionally introduced exotic marine microorganisms that can be applied to risk assessment of proposed applications of GEMBs.

4.3.3 Pathogenicity to humans

Some marine bacterial species are virulent human pathogens. Among the best known marine pathogenic bacteria are species of vibrios, including *V. cholerae*, the causative agent of cholera, *V. parahaemolyticus*, a pathogen associated with consumption of raw fish and other seafood that causes gastrointestinal disorders, and *V. vulnificus*, an opportunistic pathogen dangerous to immunosuppressed individuals (Huq and Colwell, 1995). In addition, there are bacterial species pathogenic for both marine animals and humans, e.g., *Streptococcus iniae*, which causes bacteremic illness usually accompanied by cellulitis (Anonymous, 1996b, 1996d). Certainly, other unidentified marine bacterial species exist that are pathogenic to humans.

In most cases of infectious diseases, symptoms are brought on or amplified by the action of endotoxins (which are retained within the cell) or exotoxins (which are released by the cell) produced by the pathogen. Some pathogens, such as *Pseudomonas*, are able to produce several types of toxins that can damage physiological systems in many ways, including hemolyzing blood cells, damaging neurological functions, raising body temperature, provoking loss of body fluids from the intestinal tract, and blocking protein production by the cells of the host. Most bacterial toxins are proteins whose production is controlled by a specific gene. The important point for risk assessment is that these genes may be carried within mobile genetic elements, such as plasmids, that can be transferred between bacteria living free in nature (Sörensen, 1993).

Although it is reasonable to assume that no industry would chose to develop a frank human pathogen, or transgenic variation thereof, for uncontained use in the marine environment, another issue related to pathogenicity might present problems for risk assessors. Specifically, the GEMB strain under development may be an unrecognized opportunistic pathogen; i.e., a microorganism that is easily deflected by healthy immunodefenses but is able to cause disease in immunocompromised hosts.

Powerful new therapeutic methods are being developed to treat patients suffering from cancer, diabetes, AIDS, and other chronic diseases. A major side effect of many of the treatments is that immune systems of the victims are damaged by the treatment itself so that their ability to fight off disease is impaired. These individuals are especially susceptible to infection by opportunistic pathogens. Of course, because human populations are on land, the most immediate threat to immunocompromised persons are terrestrial opportunistic pathogens. However, marine bacteria may also be a threat. For example, *V. vulnificus*, a ubiquitous marine bacterium that lives in water above 17° C, is particularly dangerous to immunocompromised persons, especially the elderly with pre-existing chronic liver disease (Booth, 1993; Hlady et al., 1993). These persons may become infected by this organism while bathing or handling shellfish or eel.

Since most marine bacteria are little known, their potential for opportunistic pathogenicity cannot be predicted, thus complicating risk assessment. Molecular biology methods are being developed to determine whether obscure bacteria are pathogenic. The first complete sequence of a bacterium, *Hemophilus influenza*, was published in 1995 (Kreeger, 1995). Since then, research has revealed significant portions or complete sequences of 31 bacteria and nine archaea (Strauss and Falkow, 1997). The entire sequence of *V. cholerae* is currently being determined. This information is being used to create assays for bacterial pathogenicity. In particular, investigators have found that certain genes associated with virulence, toxin production, adhesion, invasins, and other factors are present in pathogenic bacteria but missing in nonpathogens of the same genus or species (Mecsas and Strauss, 1996; USEPA and Agriculture and Agrifood Canada, 1997; Strauss and Falkow, 1997). Within the foreseeable future, risk assessors will be able to look for genetic features indicative of pathogenicity to humans, animals, or plants in GEMBs being considered for uncontained applications. For the present, however, ability to predict the pathogenic potential of newly developed marine bacterial species is severely limited.

Of relevance to this study is the close association existing between selected marine bacteria and other marine microorganisms. For example, an estimated 10^6 bacterial cells representing many species have been observed to adhere to egg sacks of copepod zooplankton (Huq et al., 1990; Tamplin et al., 1990). Indigenous bacterial species, such as *Moraxella* and *Pseudomonas*, have been found to directly influence the growth of, and hemolysin production by, the dinoflagellate *Amphidinium carterae* (Nayak and Karunasagar, 1997). Bacterial involvement may also influence toxin production by toxigenic dinoflagellates (Kopp et al., 1997). Vibrios and other marine bacteria living in close association with cyanobacteria, diatoms, dinoflagellates, or macroalgae may have an influence on when and where the hosts bloom (Colwell, 1996; Levasseur et al., 1996). Conditions that favor zooplankton to blooms and stimulate *V. cholerae* to revert from the VBNC phase to the culturable state. Our ability to assess whether a bacterial species released into the ocean may stimulate blooms or toxin production in microalgae, cyanobacteria, dinoflagellates, etc., is, in general, extremely limited.

5. CONCLUSION

Before transgenic microorganisms can be introduced into the marine environment—be it the open seas, coastal waters, harbors, seashores, or marine sediments—their developers will have to receive permission from appropriate federal and state regulatory agencies. Approval, more likely than not, will depend on the adequacy of data pertaining to: (1) level of familiarity with the candidate GEMB strain, (2) level of familiarity with the environment into which the candidate GEMB strain will be introduced, (3) ability of the developer to ensure containment of the GEMB, and (4) level of risk the introduced GEMB strain will pose to the environment and human health. Providing data on each of these subjects sufficient for risk analysis will, for the present, be difficult, perhaps impossible.

On the assumption that the first uncontained application of a GEMB strain will be bioremediation (although the same considerations would apply for any environmental application), several bacterial strains show promise for commercial use. Is it likely than any of the can meet the first criterion? A transgenic form of *Pseudomonas putida* is currently undergoing open-air, contained field testing on a terrestrial site, and data are being generated on the environmental effects, including the extent of interactions with wild, indigenous microorganisms. However, since *P. putida* has a long history as an opportunistic pathogen (Balows et al., 1991), several such studies will be required to demonstrate safety. As a result, it is unlikely that this organism will be sufficiently well characterized within five years to permit release into the open marine environment. Other potentially useful marine bacteria that may be modified for bioremediation or other commercial purposes are even less well understood than *P. putida*. For these reasons, no GEMB is sufficiently well characterized or well known to meet the familiarity criterion. This being the case, approval for the use of GEMBs in a field test would only be given if the organisms could be absolutely contained within the test site.

The second criterion—familiarity with the target environment—also remains an impediment to development and use of genetically engineered marine microorganisms. Knowledge of marine microbial ecology is far from complete. Most biotic and abiotic processes in the marine environment are poorly understood, and there are probably many pathways for gene transfer or other ecological effects that are as yet undiscovered. At present, the marine environment is not sufficiently well characterized to permit credible, scientific risk assessment of GEMB applications in the open marine environment. This adds weight to the conclusion stated in the foregoing paragraph; i.e., a field trial of the candidate GEMB would be permitted only if the test organisms could be contained within the test site.

The third criterion—containment—is also problematic. Physical containment of microorganisms released into the ocean, spread on a shoreline, or injected into sediment cannot be ensured. Populations may escape the site of application and establish fugitive colonies that may affect ecosystem function in unpredictable ways. Biological containment methods, including bioengineered suicide mechanisms, are under development but cannot yet guarantee control of released organisms within acceptable safety limits. Since containment cannot be assured, at present no regulatory agency would be likely to grant permission for a field test of a candidate GEMB.

Three general approaches, addressing short-term, medium-term, and long-term time constraints, may permit the development of GEMBs to continue despite scientific uncertainty. The short-term approach would apply current terrestrial regulatory practice to the marine environment. Over the last ten years, EPA has exempted from review certain small-scale field trials of genetically engineered microorganisms. Since small-scale trials in general pose little or no risk to human health or the environment, similar trials of GEMBs could also be subject to eased oversight restrictions. The familiarity criteria could be relaxed, under the condition that strict monitoring requirements be met.

Under these less strict regulatory requirements, a panel set up by EPA, including marine microbiologists, marine ecologists, oceanographers, and other experts, would review proposals for introductions on a case-by-case basis. Experimental trials in well-studied environments using familiar bacterial species that contain only innocuous foreign genes should be allowed to proceed. Techniques for monitoring GEMs in the soil environment are well developed, have proven dependable and accurate in the field (Smalla and van Elsas, 1996), and could be applied to carefully screened marine trials. Data gathered from monitoring these applications would add to knowledge of the fate of introduced organisms and genes, dispersal patterns of introduced organisms, effects of introduced organisms on sentinel animal, plant, and microbes, and impacts of introduced organisms on abiotic systems. These trials thus would generate information that would enable future developers to meet the familiarity criteria more fully and lay the groundwork for broader, commercial-scale applications.

In the medium-term, the United States should consider applying an oversight structure similar to that recently adopted in Canada. Starting September 1, 1997, a permit is required from Environment Canada before any exotic or genetically engineered microorganism can be released into the environment (Environment Canada, 1997). Environment Canada will assess risk on a case-by-case basis, considering equally the likely impact on the local microbial ecology of gene flow from exotic or genetically engineered strains.

Implementation of this process in the United States would rationalize the current illogical regulatory framework. At present, USEPA has no authority to regulate uncontained applications of blends of naturally occurring microorganisms (except if used as pesticides); genetically engineered microorganisms, on the other hand, are subject to strict control. Accordingly, an industry applying exotic—though naturally occurring—organisms for bioremediation is introducing a multitude of entire genomes, each comprised of 5,000 or more genes, into the new environment, essentially without controls. In the case of GEMMs, the presence of a single transferred gene requires the applicant to bear the full weight of TSCA regulations. Since in either case nonindigenous genomes are being introduced into novel environments, the risk, and the regulatory response, should be parallel.

By harmonizing regulations pertaining to the release of GEMMs and exotic, naturally occurring microorganisms, several benefits would result. Because the issue of possible adverse consequences of introductions of microorganisms would have to be addressed, such a regulatory shift would stimulate government-supported research into marine microbial ecology. In addition, there would be new incentives to develop a system that would enable investigators to monitor the fate of introduced microorganisms and their genes. These developments would contribute by generating data that would help fill in gaps in the science knowledge base that presently limit the growth of some sectors of the domestic marine biotechnology industry.

Over the long-term, industry and government should commit to a substantial increase in funding for basic research in all disciplines that bear on the marine environment. In the final analysis, only basic research can provide the data required to meet the familiarity criteria and support credible risk assessment. A possible starting point for a program of basic research would be the investigation of past accidental introductions of exotic marine microorganisms to determine local ecological effects. The promise of substantial economic and environmental benefits from bioremediation and other commercial applications of genetically engineered marine microorganisms justifies the investment in research efforts, the findings of which will encourage development in marine biotechnology.

REFERENCES

Adams, M.W.W. 1995. Enzymes from microorganisms in extreme environments. Chemical and Engineering News 73(51):32-42.

Althoff, K., C. Schütt, R. Steffen, R. Batel and W.E.G. Müller. 1998. Evidence of s symbiosis between bacteria of the genus *Rhodobacter* and the marine sponge *Halichondria panicea*: harbor also for putatively toxic bacteria? Marine Biology 130(3):529-536.

American Petroleum Institute (API). 1995. In Situ and On-Site Biodegradation of Refined and Fuel Oils: A Review of Technical Literature 1988-1991. Washington, DC: American Petroleum Institute.

Anonymous. 1994. Viral gastroenteritis associated with consumption of raw oysters–Florida, 1993. Morbidity and Mortality Weekly Report 43(24):446-449.

Anonymous. 1995. Cyanotech signs license agreement for mosquitocide. Genetic Engineering News 15(13):33.

Anonymous. 1996a. Genes confirm Archaea's uniqueness. Science 271:1061.

Anonymous. 1996b. Invasive infection with *Streptococcus iniae*–Ontario, 1995-1996. Morbidity and Mortality Weekly Report 45(30):650-653.

Anonymous. 1996c. Natural microorganisms gobbling up toxic waste. Sea Technology 37(10):79.

Anonymous. 1996d. Fish fans beware. Science 273:1049.

Ashelford, K.E., J.C. Fry, M.J. Day, K.E. Hill, M.A. Learner, J.R. Marchesi, C.D. Perkins and A.J. Weightman. 1997. Using microcosms to study gene transfer in aquatic habitats. FEMS Microbial Ecology 23:81-94.

Atlas, R.M. 1995. Bioremediation. Chemical and Engineering News 73(14):32-42.

Atlas, R.M. 1998. Personal communication. January 22, 1998.

Awong, J., G. Bitton and G.R. Chaudry. 1990. Microcosm for assessing survival of genetically engineered microorganisms in aquatic environments. Applied and Environmental Microbiology 56:977-983.

Azam, F. 1998. Microbial control of oceanic carbon flux: The plot thickens. Science 280:694-696.

Balows, A., Hausler Jr., K.L. Herrmann, H.D. Isenberg and H.J. Shadomy. 1991. Manual of Clinical Microbiology. Washington, DC: American Society for Microbiology.

Becker, E.W. 1994. Microalgae: Biotechnology and Microbiology. New York: Cambridge University Press.

Benmansour, A. and P. de Kinkelin. 1997. Live fish vaccines: history and perspectives. Development of Biological Standards 90:279-289.

Bernan, V.S. 1995. Marine microorganisms as a source for new natural products. In: C.G. Lundin and R.A. Zilinskas (eds.). Marine Biotechnology in the Asian Pacific Region. 44-60. Stockholm, Sweden: Sida, Marine Science Program, SAREC.

Booth, W. 1993. After 9 deaths in Florida, oyster industry confronts raw fear. Washington Post, April 19:A3-A4.

Bragg, J.R., R.C. Prince, J.B. Wilkinson and R.M. Atlas. 1992. Bioremediation for Shoreline Cleanup Following the 1989 Alaskan Oil Spill. Houston, TX: Exxon Company.

Broad, W.J. 1996. A spate of red tides menaces coastal seas. New York Times, August 27:C1,C5.

Burkholder, J.M. and H.B. Glasgow Jr. 1997. Trophic controls on stage transformations of a toxic ambush predator dinoflagellate. Journal of Eukaryotic Microbiology 44:200-205.

Capone, D.G., J.P. Zehr, H.W. Paerl, B. Bergman and E.J. Carpenter. 1997. *Trichodesmium*, a globally significant marine cyanobacterium. Science 276:1221-1229.

Carlton, J.T. 1992. Marine species introductions by ship's ballast water: an overview. In: M.R. DeVoe (ed.). Introductions and Transfers of Marine Species: Achieving a Balance Between Economic Development and Resource Protection. 23-26. Hilton Head Island, SC: South Carolina Sea Grant Consortium.

Carlton, J.T. 1995. Marine invasions and the preservation of coastal diversity. Endangered Species Update 12(4/5):1-3.

Carmichael, W.W. 1994. The toxins of cyanobacteria. Scientific American 270(1):78-86.

Chakrabarty, A.M. 1981. U.S. Patent 4 259 444.

Chao, W.L. and R.L. Feng. 1990. Survival of genetically engineered *Escherichia coli* in natural soil and river water. Journal of Applied Bacteriology 68:319-325.

Chen, S.L. and D.B. Wilson. 1997. Construction and characterization of *Escherichia coli* genetically engineered for bioremediation of Hg^{2+} contaminated environments. Applied and Environmental Microbiology 63(6):2442-2445.

Chiura, H.X. 1997. Generalized gene transfer by virus-like particles from marine bacteria. Aquatic Microbial Ecology 13:75-83.

Ciferri, O., O. Tiboni and A.M. Sanangelantoni. 1989. The genetic manipulation of cyanobacteria and its potential uses. In: R.C. Cresswell, T.A.V. Rees and N. Shah (eds.). Algal and Cyanobacterial Biotechnology. 239-271. New York: John Wiley and Sons.

Coelen, R.J. 1997. Special topic review: Virus diseases in aquaculture. World Journal of Microbiology and Biotechnology 13:365-366.

Colwell, R.R. 1996. Global climate and infectious disease: the cholera paradigm. Science 274:2025-2031.

Colwell, R.R. P.R. Brayton, D.J. Grimes, D.B. Roszak, S.A. Huq and L.M. Palmer. 1985. Viable but non-culturable *Vibrio cholerae* and related pathogens in the environment: implications for release of genetically engineered organisms. Bio/Technology 3:817-820.

Colwell, R.R. and A. Huq. 1994. Environmental reservoir of *Vibrio cholerae*: the causative agent of cholera. In: M.E. Wilson, R. Levins and A. Spielman. Annals of the New York Academy of Sciences:740. 44-54. New York: New York Academy of Sciences.

Committee on Ship's Ballast Operations, Marine Board, Commission on Engineering and Technical Systems, National Research Council. 1996. Stemming the Tide: Controlling the Introductions of Nonindigenous Species by Ship's Ballast Water. Washington, DC: National Academy Press.

Cooney, J.J., S. Wuertz, M.M. Doolittle, M.E. Miller, K. Henry and D. Ricca. 1993. Marine fungi: novel catalysts for bioremediation of oil spills. In R. Guerrero and C. Pedros-Alio (eds.). Trends in Microbial Ecology. 639-642. Barcelona: Spanish Society for Microbiology.

Cory, J.S. and R.S. Hails. 1997. The ecology and biosafety of baculoviruses. Current Opinion in Biotechnology 8:323-327.

Cotter, P.A. and J.F. Miller. 1996. Triggering bacterial virulence. Science 273:1183-1184.

Cutter Information Corporation. 1997. International Oil Spill Control Directory. Arlington, MA: Cutter Information Corporation.

Dahlberg, C., C. Linberg, V.L. Torsvik and M. Hermansson. 1997. Conjugative plasmids isolated from bacteria in marine environments show various degrees of homology to each other and are not closely related to well-characterized plasmids. Applied and Environmental Microbiology 63:4692-4697.

Davis, B.D. 1980. Gene transfer in bacteria. In: B.D. Davis, R. Dulbecco, H.N. Eisen and H.S. Ginsberg (eds.). Microbiology, Including Immunology and Molecular Genetics. 137-152. Philadelphia: Harper and Row.

De Meyer, C.P., R.H. Charlier, K. De Vos and B. Malherbe. 1997. *In situ* bioremediation of contaminated sediments. Sea Technology 38(1):57-59.

DeLong, E.F. 1997. Marine microbial diversity: the tip of the iceberg. Trends in Biotechnology 15:203-207.

Edwards, C. 1993. Monitoring Genetically Manipulated Microorganisms in the Environment. Chichester, UK: John Wiley and Sons.

Elhai, J. 1994. Genetic techniques appropriate for the biotechnological exploitation of cyanobacteria. Journal of Applied Phycology 6:177-186.

Environment Canada. 1997. Guidelines for the Notification and Testing of New Substances: Organisms (Pursuant to the New Substances Notification Regulations of the Canadian Environmental Protection Act). Ottawa, Canada: Environment Canada.

Epstein, P.R., T.E. Ford, C. Puccia and C.D. Possas. 1994. Marine ecosystem health: implications for public health. In: M.E. Wilson, R. Levins and A. Spielman. Annals of the New York Academy of Sciences:740. 13-23. New York: New York Academy of Sciences.

Falkow, S. 1997. What is a pathogen? ASM News 63(7):359-365.

Fenical, W. 1993. Chemical studies of marine bacteria: developing a new resource. Chemical Reviews 93(5):1673-1683.

Fenical, W. and P.R. Jensen. 1993. Marine microorganisms: a new biomedical resource. In: D.H. Attaway and O.R. Zaborsky (eds.). Marine Biotechnology, Volume 1: Pharmaceuticals and Bioactive Natural Products. 419-457. New York: Plenum Press.

Fuhrman, J.A., R.M. Wilcox, R.T. Noble and N.C. Law. 1993. Viruses in marine food webs. In: R. Guerrero and C. Pedros-Alio (eds.). Trends in Microbial Ecology. 295-302. Barcelona, Spain: Spanish Society for Microbiology.

Fulthorpe, R.R. and R.C. Wyndham. 1991. Transfer and expression of the catabolic plasmid pBRC60 in wild bacterial recipients in a freshwater ecosystem. Applied and Environmental Microbiology 57:1546-1553.

Gauthier, M.J., F. Cauvin and J. Breittmayer. 1985. Influence of salts and temperature on the transfer of mercury resistance from a marine pseudomonad to *Escherichia coli*. Applied and Environmental Microbiology 50:38-40.

Gelernter, W. and G.E. Schwab. 1993. Transgenic bacteria, viruses, algae and other microorganisms as *Bacillus thuringiensis* toxin delivery systems. In: P.F. Entwistle, J.S. Cory, M.J. Bailey and S. Higgs (eds.). *Bacillus thuringiensis*, An Environmental Biopesticide: Theory and Practice. 89-104. New York: John Wiley and Sons.

Geller, J. 1996. Molecular approaches to the study of marine biological invasions. In: J.D. Ferraris and S.R. Palumbi (eds.). Molecular Zoology: Advances, Strategies, and Protocols. 119-132. New York: Wiley-Liss.

Glasgow, H.B. Jr., J.M. Burkholder, D.E. Schmechel, P.A. Tester and P.A. Rublee. 1995. Insidious effects of a toxic estuarine dinoflagellate on fish survival and human health. Journal of Toxicology and Environmental Health 46:501-522.

Greer, J. 1997. In harm's way? The threat of toxic algae. Maryland Marine Notes July-August:1-4.

Groombridge, B. 1992. Global Diversity: Status of the Earth's Living Resources. New York: Chapman and Hall.

Hegde, A., J. Antony and S. Rao. 1996. Inactivation of viruses in aquaculture systems. INFOFISH International No. 6:40-43.

Heuer, H., D.F. Dwyer, K.N. Timmis and I. Wagner-Döbler. 1995. Efficacy in aquatic microcosms of a genetically engineered pseudomonad applicable for bioremediation. Microbial Ecology 29(2):203-220.

Hill, R.T., W.L. Straube, A.C. Palmisano, S.L. Gibson and R.R. Colwell. 1996. Distribution of sewage indicated by *Clostridium perfringens* at a deep-water disposal site after cessation of sewage disposal. Applied and Environmental Microbiology 62(5):1741-1746.

Hlady, W.G., R.C. Mullen and R.S. Hopkins. 1993. *Vibrio vulnificus* from raw oysters: leading cause of reported deaths from foodborne illness in Florida. Journal of the Florida Medical Association 80:536-538.

Hugenholtz, P. and N.R. Pace. 1996. Identifying microbial diversity in the natural environment: a molecular phylogenetic approach. Trends in Biotechnology 14:190-196.

Huq, A., R.R. Colwell, R. Rahman, A. Ali, M.A.R. Chowdhury, S. Parveen, D.A. Sack and E. Russek-Cohen. 1990. Detection of *Vibrio cholerae* 01 in the aquatic environment by fluorescent-monoclonal antibody and culture methods. Applied and Environmental Microbiology 56(8):2370-2373.

Huq, A. and R.R. Colwell. 1995. Vibrios in the marine and estuarine environments. Journal of Marine Biotechnology 3:60-63.

Jackman, S.C., H. Lee and J.T. Trevors. 1992. Survival, detection and containment of bacteria. Microbial Releases 1:125-154.

Jahans, K.L., G. Foster and E.S. Broughton. 1997. The characterisation of *Brucella* strains isolated from marine mammals. Veterinary Microbiology 57:373-382.

Jeffrey, S.W., M.R. Brown and K. Volkman. 1994. Haptophytes as feedstocks in mariculture. In: J.C. Green and B.S.C. Leadbeater (eds.). The Haptophyte Algae. 287-302. Oxford, UK: Clarendon Press.

Jeffrey, W.H., J.H. Paul, et al. 1990. Natural transformation of a marine vibrio species by plasmid DNA. Microbial Ecology 19(3):259-268.

Jensen, P.R. and W. Fenical. 1994. Strategies for the discovery of secondary metabolites from marine bacteria: ecological perspectives. Annual Review of Microbiology 48:559-584.

Johnson, L.E. and J.T. Carlton. 1996. Post-establishment spread in large-scale invasions: dispersal mechanisms for the zebra mussel *Dreissena polymorpha*. Ecology 77(6):1686-1690.

Joint Subcommittee on Agriculture, Shrimp Virus Work Group (JSA SVWG). 1997. An Evaluation of Potential Shrimp Virus Impacts on Cultured Shrimp and Wild Shrimp Populations in the Gulf of Mexico and Southeastern U.S. Atlantic Coastal Waters. Washington, DC: US National Marine Fisheries Service.

Kastner, R.. 1996. Get the jump on zebra mussel repercussions. Fish Farming News Sept/Oct:16.

Kopp, M., G.J. Doucette, M. Kodama, G. Gerdts, C. Schütt and L.K. Medlin. 1997. Phylogenetic analysis of selected toxic and non-toxic bacterial strains isolated from the toxic dinoflagellate *Alexandrium tamarense*. FEMS Microbiological Ecology 24:251-257.

Kreeger, K.Y. 1995. First completed microbial genomes signal birth of new area of study. The Scientist 9(23):14-15.

Lebaron, P., P. Bauda, M.C. Lett, Y. Duval-Iflah, P. Simonet, E. Jacq, N. Frank, B. Roux, B. Baleux, G. Faurie, J.C. Hubert, P. Normand, D. Prieur, S. Schmitt and J.C. Block. 1997. Recombinant plasmid mobilization between *E. coli* strains in seven sterile microcosms. Canadian Journal of Microbiology 43:534-540.

Lebaron, P., V. Roux, M.C. Lett and B. Baleux. 1993. Effects of pili rigidity and energy availability on conjugative plasmid transfer in aquatic environments. Microbial Releases 2:127-133.

Leffler, M. 1998. Harmful algal blooms on the move. Maryland Marine Notes July-August:4-5.

Leong, J.C., E. Anderson, L.M. Bootland, P.W. Chiou, M. Johnson, C. Kim, D. Mourich and G. Trobridge. 1997. Fish vaccine antigens produced or delivered by recombinant DNA technologies. Development of Biological Standards 90:267-277.

Lester, J.L. 1992. Marine species introductions and native species vitality: genetic consequences of marine introductions. In: M.R. DeVoe (ed.). Introductions and Transfers of Marine Species: Achieving a Balance Between Economic Development and Resource Protection. 23-26. Hilton Head Island, SC: South Carolina Sea Grant Consortium.

Levasseur, M., S. Michaud, J.C. Egge, J.C. Nejstgaard, R. Sanders, E. Fernandez, P.T. Solberg, B. Heimdal and M. Gosselin. 1996. Production of DMSP and DMS during a

mesocosm study of an *Emiliana huxleyi* bloom: influence of bacteria and *Calanus finmarchicus* grazing. Marine Biology 126(4):609-618.
Levy, S.B. and B.M. Marshall. 1988. Genetic transfer in the natural environment. In M. Sussman, C.H. Collins, F.A. Skinner and D.E. Stewart-Tull (eds.). The Release of Genetically-Engineered Micro-organisms. 61-76. London: Academic Press Limited.
Lorenz, M.G. and W. Wackernagel. 1993. Bacterial gene transfer in the environment. In: K. Wöhrmann and J. Tomiuk (eds.). Transgenic Organisms: Risk Assessment of Deliberate Release. 43-64. Boston: Birkhauser Verlag.
Lorenz, M.G. and W. Wackernagel. 1994. Bacterial gene transfer by natural genetic transformation in the environment. Microbiological Reviews 58(3):563-602.
MacLeod, R.A. 1965. The question of the existence of specific marine bacteria. Bacteriological Reviews 29(1):9-23.
Massana, R., A.E. Murray, C.M. Preston and E.F. DeLong. 1997. Vertical distribution and phylogenetic characterization of marine planktonic Archaea in the Santa Barbara Channel. Applied and Environmental Microbiology 63(1):50-56.
Mecsas, J. and E.J. Strauss. 1996. Molecular mechanisms of bacterial virulence: Type III secretion and pathogenicity islands. Emerging Infectious Diseases 2(4):271-285.
McClung, G. 1997. Personal communication, October 3, 1997.
Meadows, M.P. 1993. *Bacillus thuringiensis* in the environment: ecology and risk assessment. In: P.F. Entwistle, J.S. Cory, M.J. Bailey and S. Higgs (eds.). *Bacillus thuringiensis*, An Environmental Biopesticide: Theory and Practice. 193-220. New York: John Wiley and Sons.
Meints, R.H. and J.L. van Etten. 1992. Eukaryotic algae. In: M. Levin, R.J. Seidler and M. Rogul (eds.). Microbial Ecology: Principles, Methods, and Applications. 883-889. New York: McGraw-Hill, Inc.
Mhiri, C. and N.T. De Marsac. 1997. Bioremediation of sites polluted by commercial PCBs: problematical questions and perspectives. Bulletin Institute Pasteur 95(1):3-28.
Miller, R.V. and G.S. Sayler. 1992. Bacteriophage-host interactions in aquatic systems. In: E.M.H. Wellington and J.D. van Elsas (eds.). Genetic Interactions Among Microorganisms in the Natural Environment. 176-193. New York: Pergamon Press.
Mitchell, A.J. 1996. Blue-green algae. Aquaculture Magazine 22(2):81-84.
Morgan, J.A.W., G. Rhodes R.W. Pickup, C. Winstanley and J.R. Saunders. 1992. The effect of microcosm design on the survival of recombinant *Pseudomonas putida* in lake water. Microbial Releases 1:155-159.
National Research Council (NRC). 1989. Field Testing Genetically Modified Organisms: Framework for Decisions. Washington, DC: National Academy Press.
National Research Council, Committee on Molecular Marine Biology (NRC CMMB). 1994. Molecular Biology in Marine Science: Scientific Questions, Technological Approaches, and Practical Implications. Washington, DC: National Academy Press.
Nayak, B.B. and I. Karunasagar. 1997. Influence of bacteria on growth and hemolysin production by the marine dinoflagellate *Amphidinium carterae*. Marine Biology 130:35-39.
Newmand, S.G. 1993. Immunisation of fish and crustaceans. INFOFISH International No. 3:39-44.
O'Morchoe, S.B., O. Ogunseitan, G.S. Sayler and R.V. Miller. 1988. Conjugal transfer of R68.45 and FP5 between *Pseudomonas aeruginosa* strains in a freshwater environment. Applied and Environmental Microbiology 54:1923-1929.

Oliver, J.D. 1993. Nonculturability and resuscitation of *Vibrio vulnificus*. In: R. Guerrero and C. Pedros-Alio (eds.). Trends in Microbial Ecology 187-191. Barcelona, Spain: Spanish Society for Microbiology.

Olsen, G.J. 1994. Archaea, Archaea, everywhere. Nature 371:657-658.

Pace, N.R. 1997. A molecular view of microbial diversity and the biosphere. Science 276:734-740.

Patz, J.A., P.R. Epstein, T.A. Burke and J.M. Balbus. 1996. Global climate change and emerging infectious diseases. Journal of the American Medical Association 275(3):217-223.

Pickup, R.W. 1992. Detection of gene transfer in aquatic environments. In: E.M.H. Wellington and J.D. van Elsas (eds.). Genetic Interactions Among Microorganisms in the Natural Environment. 145-164. New York: Pergamon Press.

Pickup, R.W., J.A.W. Morgan and C. Winstanley. 1993. *In situ* detection of plasmid transfer in the aquatic environment. In: C. Edwards (ed.). Monitoring Genetically Manipulated Microorganisms in the Environment. 61-82. Chichester, UK: John Wiley and Sons.

Pipke, R., I. Wagner-Döbler, K.N. Timmis and D.F. Dwyer. 1992. Survival and function of a genetically engineered pseudomonad in aquatic sediment microcosms. Applied and Environmental Microbiology 58:1259-1265.

Potera, C. 1996. Biofilms invade microbiology. Science 273:1795-1797.

Prieto-Samsónov, D.L., R.I. Vázquez-Padrón, C. Ayra-Pardo, J. González-Cabrera and G.A. de la Riva. 1997. *Bacillus thuringiensis*: from biodiversity to biotechnology. Journal of Industrial Microbiology and Biotechnology 19:202-219.

Prince, R.C. 1997. Bioremediation of marine oil spills. Trends in Biotechnology 15(5):158-160.

Proctor, L.M. 1997. Advances in the study of marine viruses. Microscopy Research Techniques 37(2):136-161.

Proctor, L.M. and J.A. Fuhrman. 1990. Viral mortality of marine bacteria and cyanobacteria. Nature 343:60-62.

Radmer, R.J. 1996. Algal diversity and commercial algal products. BioScience 46(4):263-270.

Relman, D.A. and D.H. Persing. 1996. Genotypic methods for microbial identification. In: D.H. Persing (ed.). PCR Protocols for Emerging Infectious Diseases. 3-31. Washington, DC: ASM Press.

Sandaa, R. 1993. Transfer and maintenance of the plasmid RP4 in marine sediments. Microbial Releases 2:115-119.

Santavy, D.L. 1995. The diversity of microorganisms associated with marine invertebrates and their roles in the maintenance of ecosystems. In: D. Allsopp, R.R. Colwell and D.L. Hawksworth (eds.). Microbial Diversity and Ecosystem Function. 211-229. Wallingford, UK: CAB International.

Sayler, G.S. 1995. TSCA application. Field release of genetically engineered bioluminescent reporter bacteria for PAH bioremediation in subsurface soil.

Sayler, G.S., J.T. Fleming, B. Applegate and C. Werner. 1992. Nucleic acid extractions and analysis: detecting genes and their activity in the environment. In: E.M.H. Wellington and J.D. van Elsas (eds.). Genetic Interactions Among Microorganisms in the Natural Environment. 237-257. New York: Pergamon Press.

Sayler, G.S. and P. Sayre. 1996. Risk assessment for recombinant *Pseudomonads* released into the environment for hazardous waste degradation.

Scanferlato, V.S., D.R. Orvos, J.C. Cairns Jr. and G.H. Lacy. 1989. Genetically engineered *Erwinia carotovora* in aquatic microcosms: survival and effects on functional groups of indigenous bacteria. Applied and Environmental Microbiology 55:1477-1482.

Schlimme, W., M. Marchiani, K. Hanselmann and B. Jenni. 1997. Gene transfer between bacteria within digestive vacuoles of protozoa. FEMS Microbiological Ecology 23(3):239-247.

Schmidt, T.M., E.F. DeLong and N.R. Pace. 1991. Analysis of a marine picoplankton community by 16S rRNA gene cloning and sequencing. Journal of Bacteriology 173:4371-4378.

Schmitz, F.J. 1994. Cytotoxic compounds from sponges and associated microfauna. In: R.W.M. van Soest, T.M.G. van Kempen and J.C. Braekman (eds.). Sponges in Time and Space. 485-496. Rotterdam: A.A. Balkema.

Schneegurt, M.A. 1997. Personal communication, August 20, 1997.

Schweder, T. 1996 The two strategies of biological containment of genetically engineered bacteria. In: J. Tomiuk, K. Wöhrmann and A. Sentker (eds.). Transgenic Organisms: Biological and Social Implications. 113-125. Boston: Birkhauser Verlag.

Singleton, F.L., R. Attwell, S. Jangi and R.R. Colwell. 1982. Effects of temperature and salinity on *V. cholerae* growth. Applied Environmental Microbiology 44:1047-1058.

Skujins, J. 1992. Environmental control of cyanobacteria. In: M. Levin, R.J. Seidler and M. Rogul (eds.). Microbial Ecology: Principles, Methods, and Applications. 877-882. New York: McGraw-Hill, Inc.

Smalla, K. and J.D. van Elsas. 1996. Monitoring genetically modified organisms and their recombinant DNA in soil environments. In: J. Tomiuk, K. Wöhrmann and A. Sentker (eds.). Transgenic Organisms: Biological and Social Implications. 127-146. Boston: Birkhauser Verlag.

Smith, A.W., D.E. Skilling, N. Cherry, J.H. Mead and D.O. Matson. 1998. Calicivirus emergence from ocean reservoirs: zoonotic and interspecies movement. Emerging Infectious Diseases 4(1):13-20.

Sörensen, S.J. 1993. Transfer of plasmid RP4 from *Escherichia coli* K-12 to indigenous bacteria of seawater. Microbial Releases 2:135-141.

Squires, S. 1997. Beneath a microscope, the beach is always crowded. Washington Post, August 25:A3.

Staley, J.T., R.W. Castenholz, R.R. Colwell, J.G. Holt, M.D. Kane, N.R. Pace, A.A. Salyers and J.M. Tiedje. 1997. The Microbial World: Foundation of the Biosphere. 1-32. Washington, DC: American Academy of Microbiology.

Stent, G.S.1963. Molecular Biology of Bacterial Viruses. San Francisco, CA: W.H. Freeman and Company.

Stevens, W.K. 1996. Bay inhabitants are mostly aliens. New York Times, August 20:C1,C10.

Stewart, G.J. 1992. Transformations in the natural environments. In: E.M.H. Wellington and J.D. van Elsas (eds.). Genetic Interactions Among Microorganisms in the Natural Environment. 216-234. New York: Pergamon Press.

Strauss, E.J. and S. Falkow. 1997. Microbial pathogenesis: genomics and beyond. Science 276:707-712.

Suttle, C.A., A.M. Chan and M.T. Cottrell. 1990. Infection of phytoplankton by viruses and reduction of primary productivity. Nature 347:467-469.

Sybron Biochemical. No date. Harnessing Nature Through Biotechnology. Salem, VA: Sybron Chemicals Inc.
Szafranski, P., C.M. Mello, T. Sano, C.L. Smith, D.L. Kaplan and C.R. Cantor. 1997. A new approach for containment of microorganisms: dual control of streptavidan expression by antisense RNA and the T7 transcription system. Proceedings of the National Academy of Sciences USA 94(4):1059-1063.
Tabashnik, B.E. 1994. Evolution of resistance to *Bacillus thuringiensis*. Annual Review of Entomology 39:47-79.
Tamplin, M.L., A.L. Gauzens, A. Huq, D.A. Sack and R.R. Colwell. 1990. Attachment of *Vibrio cholerae* serogroup 01 to zooplankton and phytoplankton of Bangladesh waters. Applied and Environmental Microbiology 56(6):1977-1980.
Tamplin, M.L. and R.R. Colwell. 1986. Effects of microcosm salinity and organic substrate concentration on production of *V. cholerae* enterotoxin. Applied Environmental Microbiology 52:297-301.
Tester, P.A. 1994. Harmful marine phytoplankton and shellfish toxicity: potential consequences of climate change. In: M.E. Wilson, R. Levins and A. Spielman. Annals of the New York Academy of Sciences:740. 69-76. New York: New York Academy of Sciences.
Texas General Land Office (TGLO). 1992. *Mega Borg* Oil Spill: An Open Water Bioremediation Test. Austin, TX: Texas General Land Office.
Thomas, C.M. 1996. Bacterial diversity and the environment. Trends in Biotechnology 14:327-329.
Timmis, K.N., R.J. Steffan and R. Unterman. 1994. Designing microorganisms for the treatment of toxic wastes. Annual Review of Microbiology 48:525-557.
Tver, D.F. 1979. Ocean and Marine Dictionary. Centreville, MD: Cornell Maritime Press, Inc.
United Kingdom, Department of the Environment (UKDOE). 1995. Guidance for Experimental Releases of Genetically Modified Baculoviruses. Guidance Note 6. London: Department of the Environment.
United States Congress, Office of Technology Assessment (OTA). 1991. Bioremediation for Marine Oil Spills. Washington, DC: Office of Technology Assessment.
United States Environmental Protection Agency (EPA) and Agriculture and Agrifood Canada. 1997. Final Report. Environmental Fate and Effects Test Methods Workshop, Alexandria, Virginia, March 25-26, 1996. Alexandria, VA: SRA Technologies Inc.
Venosa, A.D., J.R. Haines and D.M. Allen. 1992. Efficacy of commercial inocula in enhancing biodegradation of weathered crude oil contaminating a Prince William Sound beach. Journal of Industrial Microbiology 10:1-11.
Venosa, A.D., J.R. Haines and B.L. Eberhart. 1997. Screening of bacterial products for their crude oil biodegradation effectiveness. Methods in Biotechnology 2:47-58.
Venosa, A.D., M.T. Suidan, B.A. Wrenn., K.L. Strohmeier, J.R. Haines, B.L. Eberhart, D. King and E. Holder. 1996. Bioremediation of an experimental oil spill on the shoreline of Delaware Bay. Environmental Science and Technology 30(5):1764-1775.
Wagner-Döbler, I., R. Pipke, K.N. Timmis and D.F. Dwyer. 1992. Evaluation of aquatic sediment microcosms and their use in assessing possible effects of introduced microorganisms on ecosystem parameters. Applied and Environmental Microbiology 58:1249-1258.

Walton, W.E. and M.S. Mulla. 1992. Impacts and fates of microbial pest-control agents in the aquatic environment. In: A. Rosenfield and R. Mann (eds.). Dispersal of Living Organisms Into Aquatic Ecosystems. 205-238. College Park, MD: Maryland Sea Grant College.

Watson, A.J. and P.S. Liss. 1998. Marine biological controls on climate via the carbon and sulphur geochemical cycles. Philosophical Transactions of the Royal Society London (Biology) 353:41-51.

Wiebe, P.H., R.C. Beardsley, D.G. Mountain and A. Bucklin. 1996. Global ocean ecosystem dynamics-initial program in Northwest Atlantic. Sea Technology 37(8):67-76.

Williams, J.G.K. and A.A. Szalay. 1983. Stable integration of foreign DNA into the chromosome of the cyanobacterium *Synechocycstic* R2. Gene 24:37-51.

Williams, N. 1997. Gram-positive bacterium sequenced. Science 277:478.

Wommack, K.E., R.T. Hill, M. Kessel, E. Russek-Cohen and R.R. Colwell. 1992. Distribution of viruses in the Chesapeake Bay. Applied and Environmental Microbiology 58:2965-2970.

Wommack, K.E., R.T. Hill, T.A. Muller and R.R. Colwell. 1998. Effects of sunlight on bacteriophage viability and structure. Applied and Environmental Microbiology 62(4):1336-1341.

Wood, A.M., N. Apelian and L. Shapiro. 1993. Novel toxic phytoplankton: A component of global change? In: R. Guerrero and C. Pedros-Alio (eds.). Trends in Microbial Ecology. 479-482. Barcelona, Spain: Spanish Society for Microbiology.

Xu, H., N. Roberts, F.L. Singleton, R.W. Attwell, D.J. Grimes and R.R. Colwell. 1982. Survival and viability of nonculturable *Escherichia coli* and *Vibrio cholerae* in estuarine and marine environment. Microbial Ecology 8:313-323.

Xu, H., N.C. Roberts, L.B. Adams, P.A. West, R.J. Siebling, A. Huq, M.I. Huq, R. Rahman and R.R. Colwell. 1984. An indirect fluorescent antibody staining procedure for detection of *Vibrio cholerae* serovar 01 cells in aquatic environmental samples. Journal of Microbial Methods 2:221-231.

Yoshikawa, K., T. Takadera, K. Adachi, M. Nishijima and H. Sano. 1997. Korormicin, a novel antibiotic specifically active against marine gram-negative bacteria, produced by a marine bacterium. Journal of Antibiotics (Tokyo) 50:949-953.

Zilinskas, R.A., R.R. Colwell, D.W. Lipton, and R.T. Hill. 1995. The Global Challenge of Marine Biotechnology: A Status Report on the United States, Japan, Australia and Norway. College Park, MD: Maryland Sea Grant College.

Zilinskas, R.A. and C.G. Lundin. 1993. Marine Biotechnology and Developing Countries. World Bank Discussion Paper 210. Washington, DC: World Bank.

Zobell, C.E. 1946. Marine Microbiology: A Monograph on Hydrobacteriology. Waltham, MA: Carnica Botanica Co.

Chapter 5

Federal and State Regulations Relevant to Uncontained Applications of Genetically Engineered Marine Organisms

SUSAN STENQUIST
University of Maryland Biotechnology Institute

1. INTRODUCTION

Genetically engineered organisms (GEOs) may be introduced into the open marine environment in ways that vary substantially in purpose and scale. An introduction could range in scope from a scientific experiment with microorganisms in coastal areas testing the effectiveness of bioremediation on localized oil spills to the release of thousands of transgenic fish into an offshore, open-ocean aquaculture facility for commercial purposes. As we have seen in previous chapters, the nature of the introduction will affect the level of risk. Factors that affect risk include the number of organisms released, their reproductive viability, and the environmental characteristics of the region of introduction. As we will see in this chapter, it is crucial that the variety of organisms, and the range of actions that may result in their introduction into the environment, be clearly defined and categorized in order to determine the level of regulatory oversight required. For example, which agency has authority in any given case will depend upon whether the introduction takes place in federal or state waters; in a containment facility or directly into the open water; by a federally funded research team or by a private company; whether the organism is the product of nonindigenous species or derived from native species; and whether the organism being considered for release is an animal, plant, or microorganism.

No single law or federal agency covers all aspects of these issues. This chapter examines the four major categories of laws that relate to this inquiry, including those that cover biotechnology, the coastal environment, wildlife, and marine aquaculture, also known as mariculture. (For a detailed discussion of federal regulations concerning biotechnology, see Chapter 1.) The discussion of each category of legislation in this chapter has two components: (1) introductory remarks and general considerations; and (2) descriptions and analyses of the respective laws as they relate to the issue of marine transgenics. Following the discussions of federal laws, all four categories of laws at the state level are also considered. Conclusions regarding the status of regulatory oversight of the introduction of GEOs into the open marine environment appear at the end of the chapter.

2. FEDERAL OVERSIGHT

2.1 The Coastal Environment

Freshwater experimental sites are typically land-locked bodies of water under private ownership and within the confines and jurisdiction of one state. By contrast, experimental sites in the marine environment are part of a larger, fluid medium that crosses coastal state and international boundaries and is generally considered public space. Questions concerning jurisdiction over the various parts of the marine environment and how this common property resource is managed make regulatory considerations for marine introductions quite different from those of freshwater situations. This section identifies which federal and state authorities have jurisdiction over the ocean areas in question and describes the difficulties associated with managing these areas in the public interest. The laws described and analyzed herein are related to the management of ocean resources and the protection of the marine environment.

The Supreme Court traditionally has interpreted the commerce clause of the U.S. Constitution to give authority to the federal government to regulate navigable waters. This right is commonly known as "navigation servitude" (Beatley et al., 1994). The U.S. Army Corps of Engineers (COE) is the arm within the federal government responsible for maintaining the navigability of the U.S. waters, including both navigable fresh and marine waters, while the Department of Commerce (DOC) is authorized to protect marine resources. In 1976, the Magnuson Fisheries Conservation and Management Act extended the U.S. exclusive fishery conservation zone to

200 miles offshore, and in 1983 President Reagan proclaimed the 200 mile zone a U.S. Exclusive Economic Zone (EEZ). This 200 mile zone encompasses 3.9 billion acres, making the ocean property of the United States over one and a half times greater than the 2.3 billion acres of land under its control (King and Jennings, 1988). According to the United Nations Law of the Sea treaty (to which the U.S. is not a signatory nation but generally adheres), all waters beyond 200 miles are international territory over which no country has sovereignty (McCoy, 1993a). The only exception to this rule is if the continental shelf extends beyond 200 miles. In this case, national sovereignty can extend to up to 350 miles offshore (Law of the Sea Convention, Article 76).

The majority of states maintain home rule over submerged lands up to three miles offshore, though some states, such as Florida and Texas, are seeking to control waters beyond the three mile limit (Beatley et al., 1994). There are certain situations requiring federal action in the national interest that allow the federal government to override the authority of states in state waters. As related to our investigation, if an introduction of transgenic organisms were to take place within the three-mile coastal zone, it would be subject to both federal and state regulation. If the introduction were to take place between three and 200 miles offshore, only appropriate federal regulations would have to be followed.

Along with jurisdiction rights comes the responsibility of the federal and state governments to manage the waters in the public interest. Coastal and open-ocean waters have traditionally been seen as common property resources. Because the oceans are common property and, in many ways, public goods, government must (1) protect marine and coastal resources for current and future generations; (2) mediate and resolve conflicts concerning competing claims on, and uses for, marine and coastal resources and space; and (3) obtain fair return to the public when submerged lands are rented for private endeavors (Beatley et al., 1994). Many federal laws enacted since the 1970s encourage the fair and responsible management of the ocean and coastal zones, including the Coastal Zone Management Act, the Outer Continental Shelf Lands Act, and the Marine Protection, Research and Sanctuaries Act. Except for the Coastal Zone Management Act, all of the new laws are singular in purpose (NRC, 1992); that is, they establish leasing procedures for mineral exploitation or regulate the dumping of materials into the marine environment, etc.

Eight federal laws dealing with the coastal environment relate directly to our inquiry. Each law, and its relevance to the issue of transgenic organisms, is discussed below.

2.1.1 Magnuson Fishery Conservation and Management Act of 1976 (Public Law 94-265; 16 USC 1801)

The relevance of this act to our inquiry is twofold. First, the act defines the extent of U.S. fishery jurisdiction to be seaward of the state three-mile territorial limit out to 200 miles offshore. Second, it establishes eight Regional Fishery Councils to oversee the management and conservation of the nation's fisheries resources within federal waters. The eight regions include New England, Mid-Atlantic, South Atlantic, Caribbean, Gulf of Mexico, Pacific, North Pacific, and Western Pacific. Each council represents the federal interests in that geographical area. The councils are overseen by the National Marine Fisheries Service (NMFS), an agency of DOC. All fishery management plans developed by these councils and subsequent amendments must be approved by the Secretary of Commerce.

When an activity is proposed within federal waters that may affect a fishery, advisement or authorization must be sought from the appropriate regional council. For example, it was necessary for the principals of a nine-square-mile sea scallop farm being established 10 miles off the coast of Massachusetts to seek permission from the New England Fishery Management Council to exclude certain fishing activities from the area of the facility (Plante, 1996a). The implication for future transgenic organism introductions is that if an introduction is going to take place in federal waters, and if there is any possibility of interfering with fisheries, federal oversight will be triggered through the regional fishery councils.

2.1.2 Coastal Zone Management Act of 1972 (CZMA) (Public Law 92-583; 16 USC 141 et. seq.)

The CZMA, which is administered by the National Oceanic and Atmospheric Administration (NOAA) within DOC, provides financial assistance to each state to develop and administer its own coastal management plan that is consistent with standards set in CZMA. The management plans must focus on the wise use of the land and water resources. The 1980 reauthorization, the CZM Improvement Act, stresses that states must also address living marine resource conservation in developing their coastal zone management plans. In response, many states are taking broader geographical and ecological approaches to developing their plans, including for example fisheries, marine mammals, and delicate marine habitat considerations (Beatley et al., 1994). The 1990 reauthorization amendments extended the coastal state consistency responsibility to the area "within or outside the coastal zone that affects any land or water use or natural resource of the coastal zone" (16 USC Sec 1456). This can include areas beyond the three-mile state territorial waters.

Once a state management program has been approved by the Secretary of Commerce, it triggers "federal consistency" provisions in Section 307(c). As originally enacted, these provisions require federal agencies, permittees, and licensees to show that their proposed developments in or directly affecting coastal zones are consistent with the state program (Findley and Farber, 1992).

There is a section of the CZMA called Special Area Management Planning. This section provides for the establishment of "comprehensive plans providing for natural resource protection and reasonable coastal-dependent economic growth containing a detailed and comprehensive statement of policies; standards and criteria to guide public and private uses of lands and waters; and mechanisms for timely implementation in specific geographic areas within the coastal zone" (Beatley et al., 1994). It was under this section of the CZMA that the Chesapeake Bay Program (CBP) was developed. This program is an agreement signed in 1983 by the governors of Maryland, Virginia, and Pennsylvania, the mayor of Washington, D.C., the chairman of the Chesapeake Bay Commission, and the administrator of the Environmental Protection Agency (EPA). Since then, Delaware and West Virginia have also become signatories. The CBP has developed policies that are of importance to our study by defining transgenic organisms as exotic species and by creating a procedure for obtaining introduction permits. (Refer to the section below on state regulations concerning wildlife for an in-depth discussion.)

The importance of CZMA to our study results from how it encourages states to manage their coastal zones. For example, states will have to decide which critical areas, including areas with zoning restrictions, shall be off limits for an experimental introduction. They will have to consider whether additional permit requirements will be required to show consistency or compliance with federal coastal zone management plans. States may differ in how they deal with these and other issues.

2.1.3 The Rivers and Harbors Act of 1899 (33 USC 401-413)

This act gives COE the authority to regulate structures or facilities over or in navigable waters (Ewart et al., 1995). Navigable waters are defined as "those waters that are subject to the ebb and flow of the tide and/or are presently used, or have been used in the past, or may be susceptible for use to transport interstate or foreign commerce" (Hightower, 1994). Before net pens or other gear can be deployed in navigable waters, the act requires a technical review, a public hearing if one is requested during the comment period, and a siting permit from the appropriate regional COE Office (Ewart et al., 1995). COE also must assess the impact of the proposed activity on existing uses of the waterway and the local environment (Ewart et al., 1995). Federal laws and executive orders require COE to solicit comments from several federal and state agencies including: EPA (Fish and Wildlife Coordination Act); the Fish and Wildlife Service (FWS) (16 USC 661 et seq., Fish and Wildlife Coordination Act, and E.O.11990); and NMFS (Fish and Wildlife Coordination Act) (Hightower, 1994). Only after these steps have been taken can COE issue a permit.

As related to our inquiry, if an introduction will involve the building or placement of any kind of structure in navigable waters then COE must be consulted beforehand and the Corps must, in turn, notify other agencies of the proposed action.

2.1.4 Fish and Wildlife Coordination Act of 1958 (FWCA) (Public Law 85-624; 16 USC 661 et seq.)

This act, amending the original act of 1934, requires that all water projects conducted or authorized by the federal government be reviewed by the appropriate federal and state agencies having jurisdiction over the wildlife resources of the proposed site. The main objective of the law is to ensure that wildlife conservation receives equal consideration with other features of any project plan. Water projects are defined as any action that impounds, diverts, or modifies a stream or other water body, or includes navigation in or drainage of the water. In freshwater environments, FWS must be consulted; in the marine environment, NMFS has the authority.

Provided the water project—or a transgenic organism introduction—requires some type of federal authorization (e.g., COE permit or discharge permit under the Clean Water Act), FWCA is the law that brings the project to the attention of NMFS and allows NMFS to comment on environmental decisions that affect living marine resources taken by other federal agencies

(Colligan, 1996). However, the other federal agencies need not comply with the NMFS recommendations; they must only consider them.

2.1.5 The National Environmental Policy Act of 1969 (NEPA) (Public Law 91-190; 42 USC 4321 et seq.)

This act, which is administered by EPA and coordinating agencies, requires environmental assessment (EA) for all major federal actions that significantly affect the quality of the human environment. It is primarily a procedural law; that is, the act requires certain steps be followed to determine who or what will be affected by the federal action. If the EA reveals that a significant impact may arise, then an environmental impact statement (EIS) is often prepared in response. The EIS is a more thorough study that includes discussion of the impacts of different action options. Neither of these reports requires that the action should necessarily be forgone should the assessment show it would cause harm to the environment. Also, neither of these reports define the risks associated with the action, or their probabilities and degrees of uncertainty. NEPA requirements are imposed only on federal agencies and any project funded or authorized by the federal government. These requirements are not imposed directly on private companies (Hoffmann, 1988); however, if an activity conducted by a private company requires any type of federal authorization (e.g., COE permit for construction in navigable waters) then an EA could be required depending upon how "major" the activity is. The biggest point of contention with this act centers around how to define "major."

One case related to our inquiry involving the marine environment and NEPA demonstrates the problem of resolving what constitutes "major." In 1991, the federal government (specifically COE) granted permission to American Norwegian Fish Farm, Inc., a mariculture company, to construct a facility 37 miles off the New England coast, clearly in federal waters. The Conservation Law Foundation (CLF) filed a lawsuit to stop the project claiming that this was a federally authorized project and no EA had been prepared before the permit was granted (Pollack, 1991). CLF also questioned whether this was the best use of public waters and if the lease fee was fair (Pollack, 1991). Apparently the federal government and the public interest group had different opinions about what constitutes "major." An EA was later performed. The project is still not underway, however, due to inadequate research on the part of the mariculture company.

Interestingly, NEPA has also been used in a second case related to our inquiry. In this case, the law was used to assess the impact of introducing chinook salmon (*Oncorhynchus tshawytasha*), a Pacific coast species, into Delaware Bay, an east coast ecosystem. An EIS was prepared because the New Jersey Division of Fish and Game was requesting funds for the introduction under the Federal Aid in Sport Fish Restoration Act, making it a federally funded action (ANSTF, 1994a). The application of NEPA in this case was unusual in that NEPA had never previously been used for a nonindigenous species introduction analysis (OTA, 1993); however, the involvement of federal funds in the action made the need for NEPA implementation quite clear. This case set a precedent for using NEPA to assess the impact of introducing nonindigenous organisms. From these two cases it appears that NEPA could play a role in transgenic organism introductions into the marine environment. That is, they demonstrate that NEPA has been applied to marine questions as well as to nonindigenous organism introductions.

2.1.6 The Clean Water Act of 1972 (CWA) (Public Law 29-500)

The CWA, which is administered by EPA with other cooperating agencies, gives EPA authority to regulate all discharges of pollutants from point and non-point sources in order to ensure the chemical, physical, and biological integrity of the nation's waters. This act contains guidelines specifically directed at regulating aquaculture discharges (Ewart et al., 1995). Aquatic animal production facilities are considered point sources of pollution, with waste heat and nutrient-enriched waters as the most common effluents. Complying with the CWA involves obtaining a National Pollution Discharge Elimination Systems (NPDES) permit (40 CFR) from the EPA. A provision under the act allows the EPA to delegate to individual states authority to issue NPDES permits to point-source dischargers located within their borders (Ewart et al., 1995).

This act would affect experimental introductions of transgenic organisms if they were carried out as part of mariculture arrangements; that is, NPDES permits will be required if transgenic fish are put in ocean pens with water discharging from the facility back into the ocean. Once the NPDES permit is required, then the federal government is pulled into the process and the FWCA require NMFS consultation.

It may be possible to extend the authority of the CWA further regarding transgenic organisms. Because the CWA includes biological materials in its definition of point-source pollutants, it is possible that the biological discharge (e.g., fecal matter and eggs) from GEOs could be considered exceptionally dangerous pollutants and be subject to more stringent oversight. The interpretation of biological pollutants could be stretched even further to include the transgenic organisms themselves, thereby making the CWA a general regulating mechanism for release of GEOs (Hoffmann, 1988). This might be overextending the legal boundaries of the act; the act has not been used in this context to date.

2.1.7 North American Wetlands Conservation Act of 1989 (FWS) (Public Law 101-233; 16 USC 4401)

The main purpose of this act is "to conserve North American wetland ecosystems and waterfowl and the other migratory birds and fish and wildlife that depend upon such habitats" (Public Law 101-233). Wetlands can occur either as landlocked, freshwater systems or as coastal zone, brackish water, estuary systems. The act actually is a continuation of wetland laws and executive orders that began 28 years earlier with the Wetlands Act of 1961. The main thrust of federal wetland legislation over the years has been to acquire lands, institute monitoring programs, and develop action plans. These acts have not dictated what kinds of activities can or cannot be carried out in wetlands under state or private ownership, but the directive for federal agencies to conduct activities in a manner consistent with wetland conservation does restrict the activities that can be carried out in federally owned wetlands. Therefore, introducing transgenic organisms into federally owned coastal mangroves and salt marshes could be restricted in order to conform to wetland conservation directives. However, the majority of coastal lands are under state jurisdiction. Based on the measures individual states have taken in response to the information revealed through the National Wetlands Inventory Project and the Priority Conservation Plan, regulations restricting activities in these areas may or may not be substantial.

2.1.8 Marine Protection, Research, and Sanctuaries Act 1972 (Public Law 92-532)

This act, which is administered by EPA and NOAA, has three distinct objectives: (1) to protect the marine environment through the regulation of ocean dumping; (2) to carry out a monitoring program to measure the effects of dumping on ocean, coastal, and Great Lakes basin waters and to support research to assess the long-range effects of pollution, over-fishing, and manmade changes in the ocean ecosystem; and (3) to identify marine areas of special significance and provide for their management (McClain, 1991). Under this third objective, 14 national marine sanctuaries have been created to date. Each of the sanctuaries has its own management plan designed to protect its resources while taking into account the needs of communities relying on the sanctuary.

Two areas of this act appear relevant to our study. The section on ocean dumping could be relevant if a transgenic organism, as biological matter, is considered a pollutant, thereby requiring a permit from EPA to put it in the water. The more relevant portion of this act is the marine sanctuary section. This is a general constraint on all marine activities in that they must comply with the management plans and zone restrictions of each sanctuary. However, because marine sanctuaries only constitute a small fraction of the marine environment, this act is not of great significance to our inquiry.

2.2 Summary of Federal Oversight of the Coastal Environment

In general, the relevance of the CZMA, wetland laws, and marine sanctuaries to our inquiry comes primarily from regulation as to the location in which an introduction of transgenic organisms can take place. The CWA and marine dumping permits could be stretched to play a role in controlling GEO introductions, but at this point it does not appear likely. Regional fishery councils, however, could have authority in transgenic organism introductions if it is found that an introduction may interfere with the nation's fisheries. It may be appropriate for the councils to take a proactive approach to GEO introductions by issuing advisories (which they sometimes do) calling for GEO introductions to be managed responsibly so as not to put the nation's fisheries at risk.

Finally, the FWCA and NEPA may in fact be the most powerful oversight tools related to the introduction of transgenic organisms into the marine environment because they bring projects to the attention of the appropriate oversight agency and require a review of a project as to its impact on the environment. The construction of any facility for introductions, whether by a federally funded scientific research team or by a private commercial company, will require a COE permit, thereby drawing the federal government into the process. Subsequently, in accordance with the FWCA, COE would have to solicit feedback from NMFS. Then, if it appears that living marine resources may be affected by the activity over and above a certain threshold, a NEPA EA may be required, and perhaps even an EIS. As noted above, however, the primary permitting agency may disregard the recommendations resulting from an assessment.

2.3 Wildlife

Wildlife can be defined as "living things that are neither human nor domesticated" (Guralnik, 1986) or, as in the Lacey Act Amendments of 1981, "any wild animals, whether alive or dead, including without limitation any wild mammal, bird, reptile, amphibian, fish, mollusk, crustacean, arthropod, coelenterate, or other invertebrate, whether or not bred, hatched, or born in captivity, and includes any part, product, egg, or offspring thereof" (U.S. Congress, 1981). Transgenic animals clearly fall under these definitions; therefore, any activity aiming to introduce them into the environment falls under the purview of wildlife statutes and regulations. This section presents those federal wildlife laws, executive orders, and international treaties that are related to our inquiry. Whether transgenic organisms should be regulated under these statutes is debated, and at the end of the section conclusions are drawn as to the effectiveness of current wildlife statutes.

The fundamental question concerning the applicability of wildlife regulations to transgenic organisms is whether or not these organisms should be defined as nonindigenous and, thereby, be subject to regulations for nonindigenous species. At the federal level, GEOs have not been officially designated "exotic" or "nonindigenous" by any law or regulation.

Biologically speaking, this question can be looked at from many angles. For example, if a wild salmon is taken from its native waters and injected with a growth hormone so that it grows to be 11 times its natural size, is it an exotic species? Undoubtedly, this salmon would have a larger gape size than its wild cousin, and could thereby prey on larger life forms (Hallerman and Kapuscinski, 1992) potentially having an entirely different ecological impact. Another case would be a wild salmon taken from its native waters and injected with new disease resistant genes. This new ability to resist disease could lead to it being favored over the wild form in the natural community, possibly enabling it to dominate and contribute to the decline of natural populations (Hallerman and Kapuscinski, 1992). Should these transgenic fish be designated as exotic species? Are the transgenic salmon described in these examples qualitatively different from salmon that are selectively bred for large size or vaccinated against disease in aquaculture facilities? Which characteristic of the GEO should determine how it is defined—its size, the one completely foreign gene, or the uncertainty about other phenotypic expressions that may result from the new gene's presence in the organism's genome? According to Regal (1994) transgenic organisms differ from conventional selectively-bred organisms in three ways: (1) they are true ecological novelties because they "leapfrogged" phylogenetically without having evolved through natural selection; (2) they obtained the new genetic modifications without having suffered the traditional, debilitating physiological and adaptive trade-offs; and (3) they may have been changed in the non-Mendelian portions of the genome, areas that traditional selective breeding does not affect. This means that the evolutionary changes made in one genetically engineered transfer are far beyond those ever possible through traditional selective breeding. The outcome could be an organism that is unfit to survive in the wild, or it could indeed be a super-organism.

These hazy biological distinctions create a regulatory nightmare. Clearly, if the parent organism is not native to the intended release environment, then its transgenic progeny should be considered an exotic species. Both the Aquatic Nuisance Species Task Force and the now defunct Office of Technology Assessment have stated that they consider all GEOs to be non-naturally occurring and with no historic range, thereby clearly fitting the definitions of nonindigenous species (ANSTF, 1994a; OTA, 1993). But categorically defining all GEOs as exotic species may create undue regulatory and economic hardships where they could be avoided. The GEO/exotic distinction is an issue that is raised several times in this book and is discussed in this chapter specifically as it relates to regulatory oversight.

Three federal laws, one executive order, and several international treaties, all dealing with wildlife, relate directly to our inquiry.

2.3.1 Lacey Act of 1900 (Administered by FWS) (Public Law 97-79; 16 USC 3371)

The Lacey Act and its subsequent amendments have two main functions: (1) to prevent the trade of illegally taken wildlife; and (2) to prevent the importation of wildlife that is potentially injurious to human health, forestry, agriculture, horticulture, or other wildlife resources of the United States (ANSTF, 1994a). Under this act, importing and introducing organisms are considered equivalent actions. For our inquiry, the most relevant portion of the Lacy Act is the second component.

Regulations dealing with injurious wildlife were adopted in 1948, the details of which are found in Title 50, Part 16 of the Code of Federal Regulations. The main thrust of these regulations is to empower the federal government to restrict the importation of injurious wildlife into the United States pursuant to a list of prohibited species. It also authorizes FWS to enforce restrictions on interstate movements of injurious wildlife based on lists of species outlawed in each state. Reliance on state lists limits the authority of the federal government. Because adding new species to the lists is a slow and arduous process, many potentially injurious species remain unlisted (ANSTF, 1994a; OTA, 1993). Transgenic organisms are not mentioned in these lists and are presumably not covered under this authority.

An important section of the injurious wildlife regulations is found in 50 CFR 16.13. It states, "No...live fish, mollusks, crustacean, or any progeny or eggs thereof may be released into the wild except by the State wildlife conservation agency having jurisdiction over the area of release or by persons having prior written permission from such agency." In other words, anyone seeking to introduce a transgenic organism into the wild must have permission from the state in which the release is to occur, no matter whether the transgenic organism is considered exotic, native, injurious, or nonthreatening. Provided the individual states have empowered their agencies to fulfill this function, states clearly have significant authority to dictate what living organisms are placed in their waters.

2.3.2 Executive Order 11987 of 1977

This order prohibits federal agencies, grantees they fund, or activities they authorize from introducing exotic organisms without first determining the level of risk. It requires federal agencies to restrict the introduction of exotic species into lands and waters under their jurisdiction, restricts importation by U.S. citizens of exotic organisms for introduction into any natural ecosystem, and restricts export by U.S. citizens of native species for introduction outside the U.S. In 1978, the FWS drafted regulations to implement this executive order and uses them as internal guidelines (Clugston, 1986); however, no government-wide guidelines have ever been established (Peoples et al., 1992).

If this executive order had been fully implemented and GEOs were legally considered exotic species, it would be highly relevant to our study, particularly in light of the fact that the federal government has jurisdiction over the majority of the U.S. marine environment. However, its authority in the present form would only be triggered if a permit of some kind were required from the FWS, the only federal agency complying with the order. FWS has very minimal jurisdiction in the open marine environment (only over marine organisms that go on to land, such as sea turtles).

2.3.3 Endangered Species Act of 1973 (Public Law 91-135; 16 USC 1531-1544)

Under this act, the Department of Interior (DOI) is responsible for overseeing land animals and freshwater fish; the Department of Commerce (DOC) is responsible for overseeing marine mammals and fish; and the Department of Agriculture (USDA) is responsible for overseeing the import or export of land plants (McClain, 1991). The purpose of this act is to protect threatened and endangered flora and fauna, including their habitats.

Any proposal to introduce transgenic organisms must show that the organisms will in no way harm an endangered species. If an endangered species could be negatively affected by the introduction of a transgenic organism, a permit for introduction almost certainly would not be issued. This act also requires that an introduced transgenic organism likely to be preyed upon by an endangered or threatened species cannot cause harm to the predator.

2.3.4 International Treaties

Several multilateral and bilateral treaties obligate the signatory parties to ensure the safety of specified flora, fauna, and habitat types (OTA, 1993). For example, the 1972 Migratory Bird Convention (24 UST 3329) between the United States and Japan requires, among other things, that the two countries take measures to control the introduction of live plants and animals that may disrupt the ecological balance of existing systems, particularly of unique island habitats (Peoples et al., 1992). Two other treaties that have similar provisions are the International Plant Protection Convention and the Convention for Protection and Development of Marine Resources of the Wider Caribbean Region. Although treaties lack the enforcement powers of national laws, their provisions are taken into account by the appropriate federal agencies before permits are issued.

The United States and Canada have a long-standing history of cooperation in environmental regulation of the marine and freshwater systems that form much of their common border. Several joint commissions, conventions, and agreements have been established pursuant to the Boundary Waters Treaty for this purpose, including the 1909 International Joint Commission (IJC), the 1955 Convention on Great Lakes Fisheries (CGLF), and the 1972 Great Lakes Water Quality Agreement (GLWQA). The IJC monitors the Great Lakes and the St Croix, Rainy, and Red River basins. CGLF was established for the expressed purpose of combating the invasion of the sea lamprey, an exotic species that entered the Lakes through canals constructed in the early part of the century. CLWQA is aimed at reducing and remediating pollution. Though neither the Boundary Waters Treaty nor the agreements that followed from its adoption specifically address biotechnology products or the marine environment, precedents have been set for bilateral treatment of animal introductions into shared waters.

In addition, for 20 years the United States and Mexico have held formalized annual talks on issues of common interest relating to the Gulf of Mexico, including the monitoring and protection of endangered species, the enhancement of fishery yields, the improvement of fishing technologies, and the development of aquaculture. Potential GEO introductions have not been addressed, but, as developments in marine biotechnology proceed, issues relating to these products will certainly arise in discussions between the two nations.

2.3.5 Nonindigenous Aquatic Nuisance Prevention and Control Act of 1990 (NANPCA) (Public Law 101-645; 16 USC 4701) and National Invasive Species Act of 1996 (Public Law 104-332)

The impetus for creating this legislation was the invasion of zebra mussels in the Great Lakes and the more general problem of nonindigenous organisms being introduced through the ballast water of ships. The 1990 act was designed specifically to address problems occurring in the Great Lakes basin. The 1996 reauthorization expanded coverage to include all territorial waters of the United States, including the offshore Exclusive Economic Zone.

Nonindigenous species are defined as, "any species or other viable biological material that enters an ecosystem beyond its historic range, including any such organism transferred from one country into another"

(Public Law 101-646, section 1003, 9). The text of the legislation does not refer to GEOs, nor was it intended to cover them. The definition of nonindigenous species, however, is general enough that it could be interpreted to include transgenic organisms.

Under the 1990 act, the Aquatic Nuisance Species Task Force (ANSTF), constituted of representatives from several federal agencies, was created to develop the ANS program. The program was completed in 1994 and implementation of its protocols began that year. It focuses specifically on the prevention of unintentional introductions, and simultaneously provides a framework for devising management plans to deal with current and future aquatic nuisance species. The ANS program defines deliberate releases and accidental aquaculture escapes as intentional introductions that are not covered under this program. Because our inquiry concerns intentional introductions, the relevance of the ANS program to our investigation is limited. That is, only if an intentionally introduced organism becomes established as a nuisance species will the program be triggered. Under these circumstances, a management plan resulting in the eradication of the organism would have to be created. This is an after-the-fact regulation that would affect planning for a transgenic organism introduction only to a minor degree (i.e., planners would need to consider the possibility of the introduced species becoming a nuisance species).

One section of the 1990 act does address intentional introductions of nonindigenous species. Section 1207 requires the ANS task force to "identify and evaluate approaches for reducing the risk of adverse consequences associated with intentional introduction of aquatic organisms." This assessment is to be done in collaboration with regional, state, and local governments and other entities affected. A final report on this subject by the ANS task force was presented to Congress in early 1994. The 1996 amendments to the NANPCA respond to portions of the report by commissioning national ecological and ballast discharge surveys, establishing voluntary guidelines to have ships exchange ballast water in international waters, and encouraging the development of technologies and practices to reduce the likelihood of incidental introductions of exotic species.

The ANS report states at the outset that the task force did not feel prepared to include transgenic organisms in its recommendations. The 1996 reauthorization continues the practice of sidestepping issues relating to biotechnology products. However, in light of the fact that transgenic organisms have no "historic range," any placement of GEOs into a natural ecosystem would constitute a de facto introduction of a nonindigenous species according to the definitions adopted by the task force (ANSTF, 1994a).

The report makes its recommendations based on two underlying conclusions: (1) that decisions on introductions should be based on ecosystem considerations and ecological and evolutionary history (as opposed to jurisdictional boundaries); and (2) that recommendations should apply only to new introductions, not to species that have already become established in non-native habitats. The ANS task force believes prevention is the key to risk reduction, therefore most recommendations center around the decision-making process (Lassuy, 1994). The report includes recommendations to bolster existing authorities; that is, it suggests that federal agencies more fully employ NEPA, Executive Order 11987, and the Lacey Act. It also recommends that a joint federal USDA, APHIS, FWS, and NMFS permitting system be instituted with state, federal, and interjurisdictional panels to review proposed introductions (Lassuy, 1994).

As mentioned in the report, the ANS task force encountered strong opposition from several state agencies and private industries that rely heavily on the use of nonindigenous species. The activities of these entities include sport fish stocking for recreational fishing, aquaculture, the ornamental fish trade, pest control, and more (ANSTF, 1994a). The final report, however, does not contain recommendations to ban any activities; rather, it contains recommendations that emphasize risk analysis and preventative measures. To facilitate this preventative approach, the Risk Assessment and Management Committee of the ANS task force prepared the Generic Nonindigenous Aquatic Organisms Risk Analysis Review Process. This review process provides managers with a protocol for estimating the risk associated with nonindigenous aquatic organism introductions and options for managing that risk. If used faithfully, a protocol of this sort could go a long way in identifying problems and risks associated with introductions before they are manifested. It could also serve as an appropriate protocol for transgenic organism introduction reviews.

As mentioned above, as long as GEOs are not specifically included when nonindigenous species are defined, the only part of this legislation that could have direct bearing on our inquiry would be through the requirement for eradication plans if an introduced organism were to become classified a nuisance under the provisions of the law.

Three federal acts relate to plants as wildlife. These laws, specific to controlling plant importation and transportation, are the same acts being used as authorities to control genetically engineered plants under the

Coordinated Framework. These include the Plant Quarantine Act of 1912, the Federal Plant Pest Act of 1957, and the Federal Noxious Weed Act of 1974. These laws will not be discussed in detail.

2.4 Summary of Federal Oversight of Wildlife

At this time, there is no official legislation or regulation that explicitly defines genetically engineered animals as exotic species, thereby legally binding them to nonindigenous species legislation. And, there is no direct federal oversight on their interstate transport or release (OTA, 1993). This being the case, transgenic animal introductions are virtually unregulated at the federal level. It should be noted, however, that not all finfish or shellfish GEO introductions would slip through federal jurisdiction. Those organisms intended for human consumption would fall under FDA oversight (as discussed in the next section). In this case, even though FDA would be required to assess risk to the environment from an introduction, FDA would be making this assessment out of its area of expertise and would have no over-arching national policy for guidance on the subject of introduced species.

If we assume for a moment that transgenic animals could be officially incorporated into regulatory definitions of nonindigenous species, they would still not have adequate oversight; many shortcomings in the federal oversight of nonindigenous organism introductions would carry over to GEO introductions as well. For instance, despite the existence of legislation at the federal and state level, a 1993 OTA study found that approximately 1/2 of all intentionally introduced fish and mollusks have had harmful economic or environmental effects. Similarly, the study found that approximately 2/3 of all fish and 1/2 of all mollusks unintentionally introduced have had harmful economic or environmental effects. Based on these findings, OTA concluded that intentional introductions are as likely to be harmful as unintentional ones. What these figures indicate is that past introductions of nonindigenous species have been ignored or undermanaged by legislative and regulating authorities. Not only must more rigorous risk assessments be required and conducted before any introduction is permitted, but authorities must be able to enforce existing regulations and laws, and of course be adequately funded.

In the absence of a national policy on the subject of non-indigenous introductions and a legal definition of GEOs as exotic organisms, by default, regulation of transgenic animal introductions up to three miles off shore devolves to individual states. In the case of marine organism introductions, this is not the optimal approach. As noted in Chapters 2, 3, and 4, the characteristics of the marine environment dictate that introductions be considered a regional issue. This situation is similar to issues relating to clean water and clean air, both of which are under federal guidance. A discussion of how coastal states have addressed the issue of transgenic organism introductions appears later in this chapter.

2.5 Marine Aquaculture

Significant advances are being achieved in marine biotechnology research, including the development of transgenic fish and shellfish that soon may be suitable for introduction into the marine environment. Examples of transgenic fish work include growth-enhanced abalone at Stanford University in California (Powers, 1995) and cold-tolerant salmon being developed by Aqua Bounty Farms in Massachusetts (CNI, 1995). Both animals ultimately may find use in mariculture. In the more immediate future, developers of transgenic fish are likely to undertake controlled, small-scale trial introductions. Such trials probably will require the construction of containment facilities. These are likely be floating pens or concrete structures placed in public, navigable waters of the ocean. If so, researchers will have to make certain they are in compliance with the mariculture laws and regulations relevant to trial introductions. The current hurdles that mariculturists face include (1) obtaining permits to build or place facilities in public waters; (2) meeting requirements to obtain waste water discharge permits; (3) convincing authorities of the benefits of using a nonindigenous species; and (4) completing adequate plans to prevent accidental escapes of cultivated animals (NRC, 1992).

In addition, since mariculture products are intended for use as a food source for humans, many aspects of the oversight of genetically engineered fish and shellfish fall under the purview of the Food and Drug Administration (FDA). Companies intending to develop and market such products will be subject to food safety regulations.

This section presents a discussion of federal aquaculture laws, the current regulatory status relating to mariculture, the role of food safety regulations, and a review of how oversight uncertainties associated with this new industry may affect plans to introduce transgenic organisms into the marine environment.

One federal law relates specifically to aquaculture. A second targets food safety. Several other bills relevant to our inquiry are pending in Congress.

2.5.1 The National Aquaculture Act of 1980 (16 USC 2801-2810)

This act is a declaration of the great potential that aquaculture offers for augmenting existing commercial and recreational fisheries. In view of that potential, the act states that the national policy is to encourage the development of aquaculture in the U.S. It specifies the forming of a Joint Subcommittee on Aquaculture (JSA) to coordinate and encourage federal aquaculture initiatives, though no funds have ever been appropriated to support the plan developed by JSA in 1983 (Johnson, 1995). The subsequent National Aquaculture Improvement Act of 1985 appointed USDA as the lead federal agency to coordinate and disseminate aquaculture information, but it did not appoint USDA as the ultimate leader of the industry. In fact, the USDA, DOC, and DOI all have jurisdiction over the industry and continue to debate their respective roles (Stevens, 1996a). For now it appears that each federal agency is pursuing its own marine aquaculture policy (NRC, 1992).

The National Aquaculture Act is up for reauthorization in 1997. As of publication Congress still had not reauthorized it (Stevens, 1996b) leaving JSA without legislative and regulatory authority (Stevens, 1996c). However, JSA is an official subcommittee of the President's National Science and Technology Council, enabling it to continue its efforts. Through a series of regional meetings with aquaculture leaders coordinated by JSA, a new version of the National Aquaculture Development Plan is being drafted. The new plan will attempt to coordinate the agencies that oversee aquaculture and redefine the federal government's role. It was presented to Congress at the end of 1996.

Regarding our inquiry, this act is only relevant in that it addresses federal oversight of aquaculture, one of the possible commercial outlets for transgenic fish. It also appoints USDA as the federal agency responsible for disseminating aquaculture information, a position that USDA could possibly use to issue advisories on the risks associated with using transgenic fish in aquaculture and suggest appropriate risk assessment protocols. Also, a newly authorized version of the National Aquaculture Act or the newly drafted National Aquaculture Development Plan could be sources for policies regarding the use of transgenic organisms in aquaculture.

2.5.2 Current marine aquaculture bills

There were several bills introduced in Congress in the mid-1990s related to marine aquaculture, including: the Marine Aquaculture Act of 1995 and the National Aquaculture Development, Research, and Promotion Act, both in the Senate; and the Aquaculture Employment Investment Act in the House of Representatives. Each bill presents a different format for allocating regulatory authority among either USDA or NOAA. The first bill even has a section devoted to biotechnology in aquaculture (Section 4) and addresses the use of nonindigenous species and the need for environmental safeguards to protect wild fish stocks and the environment. Overall, it is apparent that mariculture is becoming a more prominent issue in Congress. Perhaps new marine aquaculture legislation could include sections specific to transgenic organisms in aquaculture and provide statutory authority for requiring some type of risk analysis before a GEO is introduced and cultivated in the open ocean.

2.5.3 Federal Food, Drug and Cosmetic Act of 1993 (FFDCA) (21 USC 500-599)

This legislation charges FDA with oversight of food and food additives intended for human consumption, and of drugs, feed, and food additives for animals. Since no legislation specifically targets biotechnology processes (see Chapter 1 for a review of regulatory history), FDA assesses a food animal with an inserted foreign gene that expresses the production of growth hormone, to give one example, the same way it would an animal that has the hormone administered through feed, injection, or other conventional means. The unit of FDA responsible for enforcing regulations relating to drugs administered to animals is the Center for Veterinary Medicine (CVM).

In practice, once regulatory authority has been delegated to an agency, risk assessors at that agency often assume responsibility for all aspects of the product's development. Therefore, CVM, through its responsibility for oversight of animal drugs, is involved with evaluating issues relating to food safety, security during research and development, and the potential ecological risks that may arise from transgenic fish and shellfish.

2.5.4 Summary of Federal Oversight of Marine Aquaculture

The location of any introduction facility will dictate the level of regulatory oversight. Onshore or nearshore facilities fall under state authority because they are within three miles of shore. A facility in open-ocean waters would be subject to federal oversight. At this point, however, there is no overarching federal policy on marine aquaculture (NRC, 1992), and there are no provisions in place to facilitate the leasing of federal waters for this purpose. Nevertheless, new precedents have been set in this arena during the past year. Having learned from the mistakes of the failed attempt to establish a salmon facility by the American Norwegian Fish Farm, Inc., project planners of a new experimental sea scallop farm ten miles off the tip of Martha's Vineyard have successfully met all regulatory requirements. They have obtained appropriate permits and authorizations from COE and the New England Fishery Management Council (Plante, 1996a) and their operation is now underway. Additionally, an experimental tuna facility has been set up in federal waters off of Virginia (Plante, 1996b). The approaches used by both of these projects will serve as models for similar ventures in the future.

Recent activity relating to marine aquaculture in Congress indicates that efforts are being made to address the deficiencies in federal guidance and oversight on the issue. As mentioned earlier, this could be an opportunity to include specific references to transgenic organism introductions in any new legislation.

Aside from the lack of a national policy on mariculture, other obstacles to the development of open ocean aquaculture are related to physical and financial constraints associated with construction of facilities (McCoy, 1993a, 1993b). Cage and servicing costs, regulatory costs and time delays, and engineering quandaries have all discouraged open-ocean enterprises. Experts advise that any marine facility must be built to withstand the worst storm (McCoy, 1993a). Until recently, technology was not available to meet this standard. Advances are being made in this arena—an experimental sea cage anchored 24 miles off of Virginia's Eastern Shore survived Bertha, one of the first hurricanes of the 1996 season to hit the U.S. coast (Plante, 1996b).

The issue of engineering constraints is particularly worrisome when considering the use of nonindigenous species or transgenic organisms in mariculture. As stated in the ANS Task Force Report on Intentionally Introduced Species, and supported by various studies (Courtenay and Williams, 1992), escapes from aquaculture facilities are virtually inevitable in any case. Adding stormy marine waters and unproven holding pens to the equation guarantees one result—transgenic organisms used in field trials or for mariculture will escape at some point. If transgenic organisms are going to be field tested or otherwise utilized in mariculture facilities, great advances must be made in holding pen construction and in creating transgenic fish certain to remain sterile should they escape.

The issue of inevitable escape leads us back to the first paragraph of this chapter; that is, who has regulatory oversight? Individually, states have tackled the leasing issue, which is appropriate since lease sites are stationary within their waters. But, because the marine environment is fluid and not restricted by the boundaries of states, the issue of introducing new organisms, either nonindigenous or transgenic, is a regional issue requiring regional collaboration. In the absence of federal guidance on the subjects of mariculture and introduced species, regional fisheries commissions could develop guidelines and advisories to aid in decision making at the state level and to provide the basis for regional policy making. The Chesapeake Bay Program's Policy for the Introduction of Non-Indigenous Aquatic Species is one such example; it will be discussed further in the next section.

The lack of clarity in the assignment of regulatory authority is particularly apparent in the fact that CVM, by default, has come to have responsibility for ecological risk assessment associated with transgenic fish and shellfish. Even within the agency, FDA has not formally detailed how FFDCA applies to regulation of genetically engineered food animals. This deficiency may be addressed at some point in the future when and if more biotechnology product applications are submitted. At present, transgenic fish and shellfish represent a minuscule percentage of FDA's concerns (Matheson, 1997).

3. STATE OVERSIGHT

As demonstrated by the foregoing section on federal interests pertaining to the marine environment, substantial responsibility for oversight and management lies with individual states and localities (Beatley et al., 1994). In the course of this inquiry, information was solicited from all 23 coastal states to determine how they regulate biotechnology, wildlife, and aquaculture/mariculture. Sixteen states responded with information that was of sufficient quality to be included in this study (see Appendix). A discussion and analysis of the information provided by the states through early 1996 is presented in this section.

3.1 Biotechnology

During the middle 1970s, in response to public concerns about experiments that employed recombinant DNA (rDNA) techniques, the National Institute of Health (NIH) drafted rDNA research guidelines. Subsequently, some states adopted their own statutes to control research within their borders. For example, the state legislatures of Maryland and New York mandated compliance with NIH guidelines—though Maryland's act expired under a sunset clause in 1982 (Hoffmann, 1988). Further, a few individual townships also took statutory action during this period, particularly townships in Massachusetts, New Jersey, and California that hosted research centers in which rDNA research was performed. Most local ordinances took the form of official adoption of the NIH guidelines, but some included provisions that were more restrictive than federal guidelines (Hoffmann, 1988). Two townships in New Jersey in 1988 went so far as to restrict the outdoor testing of genetically engineered organisms. These ordinances required research firms to carry $5 million in liability insurance and to endure several rounds of public hearings before a permit for experimentation would be issued (Gladwell, 1988). Later, the trend toward subjecting rDNA research to local legal restrictions began to reverse itself. In 1994, for example, Virginia passed the Biotechnology Research Act that prevented localities from passing any laws or ordinances that would restrict biotechnology research in their jurisdictions (Farmer and Buniva, 1995).

The history of biotechnology research legislation shows that interest in the subject has been periodic and localized. Only during the first wave of rDNA research did the issue receive much attention, and it was only in those locations where the research was being conducted that it was subjected to restrictions. Our review of state legislation on biotechnology reveals that this pattern continues. When coastal states were asked if they had instituted regulations on experimental releases of genetically engineered organisms, the only affirmative answers came from states that participate in the CBP (refer to the section on state wildlife regulations) and Mississippi, which has addressed this issue for aquaculture exclusively. North Carolina had legislation on experimental releases, but it was allowed to sunset in 1995. Three states indicated that they are considering the subject and may adopt policies in the near future. Several states without marine coastal waters, including Illinois, Minnesota, Oklahoma, and

Wisconsin, have legislation regulating field testing of genetically manipulated organisms (NBIAP, 1995). When the question was broadened, and states were asked what agency within their governments would regulate biotechnology, most states responded with "Department of Agriculture," "various," "?," or "none." States may be following the federal model and using various agencies to regulate biotechnology, they may still be grappling with the new subject of GEOs and deciding how to assign regulatory authority, or they may not yet be considering the issue.

California is an interesting case study on the topic of transgenic introductions. It is the home of two activist townships, Berkeley and Emeryville, that instituted their own rDNA statutes in the 1970s (Hoffmann, 1988). At that time there also were strong efforts to enact statewide legislation requiring California researchers to follow NIH guidelines, though these efforts ultimately did not result in legislation. In 1984, however, the Assembly Concurrent Resolution 170 was adopted which specifies that the state shall "promote the biotechnology industry, while at the same time protect public health and safety and the environment" (Hoffmann, 1988). As called for in the resolution, the California Interagency Task Force on Biotechnology was established and charged with assessing the state's legislative and regulatory framework for its ability to address biotechnology issues. The task force concluded that the state already had enough regulations to cover biotechnology and prepared a guidebook to explain which agencies, regulatory programs, and procedures must be followed for each type of biotechnology product (ITFB, 1986). When surveyed, sources at the California Department of Fish and Game stated that it may develop a policy on GEO releases in the near future. The state currently is allowing an experimental introduction of transgenic abalone in a Pacific mariculture facility under the direction of Dr. Dennis Powers of Stanford University. The conditions for this experiment are laid out in a memorandum of understanding between the California Department of Fish and Game and Stanford University (Powers, 1995). The MOU states that the California Department of Fish and Game has inspected the facilities and deemed them adequate to prevent escape of the abalone. It also mandates that at the end of the experiment all research animals be destroyed or sterilized.

3.2 The Coastal Environment

By 1994, approximately 31 states, including the Great Lakes states, had created coastal zone management plans in response to the federal CZMA (Beatley et al., 1994). These plans manage activities permitted within the states' coastal zones and mandate the types of permits required to conduct these activities. For example, attempts to introduce transgenic organisms into coastal zones may encounter facility location restrictions and water discharge requirements as mandated by state plans. The state of Hawaii, for example, has many miles of zoned coastline in which no discharges of any kind are allowed, creating a de facto prohibition on aquaculture facilities in general (NRC, 1992) and facilities necessary for the experimental introduction of transgenic organisms in particular.

Our state survey revealed that there is little consistency in marine resource oversight at the state level (see Appendix). For example, Maine and Mississippi have entire departments established specifically to address marine resources while Texas deals with marine resources through its Parks and Wildlife Department.

Most coastal states are members of one of three regional marine fisheries commissions that develop fishery management plans and issue advisories concerning the marine environment in the respective regions. In a sense, these three commissions, serving the Atlantic, Gulf, and Pacific states, are counterparts to the federal regional fishery councils as established under the Magnuson Act. These commissions oversee fishery activities within state territories in the way councils are in charge of federal waters. However, with the exception of the Atlantic States Marine Fishery Commission, which has been granted significant authority under the Atlantic Coastal Fisheries Cooperative Management Act of 1993, the commissions do not have the same legal authority as the federal councils do. The significance of these regional commissions to our inquiry is that they could act as a source of guidance on the subject of transgenic marine organism introductions by issuing advisories to the states, including suggesting appropriate risk assessment protocols.

3.3 Wildlife

Aside from the federal Lacey Act restrictions, states retain virtually all authority over decisions to import or release fish and wildlife within their jurisdictions (OTA, 1993). This is in contrast to the extensive federal guidance over the release of weeds and other plant pests (OTA, 1993).

With respect to fish and wildlife, our survey of states revealed that 14 states have specific legislation and/or regulations that restrict the introduction of nonindigenous species into their waters. Four states—California, Connecticut, Louisiana, and Texas—have a list of species that may not be imported. These states and six others—Georgia, Maine, Maryland, Mississippi, New York, and South Carolina—also require permits for all non-native introductions. Florida prohibits all exotic organism introductions into the marine environment. Virginia handles the issue a bit differently; it has issued a list of acceptable exotic species, but requires permits for all others. Washington requires that an assessment of the potential environmental risks and benefits be done before an introduction can take place. Massachusetts prohibits all freshwater introductions but has not yet addressed the issue of exotic marine introductions. Finally, though Rhode Island does not have specific legislation on the topic, state officials claim that exotic species cannot be introduced into state waters (Ganz, 1995).

Since most of the surveyed states do have legislation containing provisions regarding permitted and prohibited exotic species introductions, the main question remaining is what is the criteria state officials will use to make decisions regarding the granting of permits to introduce GEOs, particularly those types of organisms not specifically mentioned in the regulations. These states also need to consider which risk assessment protocols they will use. Approaches vary from state to state. Washington, for example, requires that a proposed introduction be accompanied by an environmental assessment as mandated by the State Environmental Protection Act. This assessment is similar to that required by NEPA for major federal actions. California has a concise list of requirements that the proposed introduction must abide by before a permit is issued. These mandate that (1) the potential impact of the proposed organism on native species must be negligible or positive; (2) the initial introduction must be reversible; and (3) a clear need for the introduction must be demonstrated. Even though these two states spell out the requirements, it is unclear how they assess the potential impact of an introduction; that is, which risk assessment methodology is used.

Another important issue for states is whether transgenic organisms should be regulated as nonindigenous organisms. When asked whether they had specifically defined GEOs as exotic species, only five states answered affirmatively. Three of these states have not defined them as exotic in any legislation or regulatory order; rather they are parties to the CBP's policy on the introduction of nonindigenous aquatic species, which specifically defines genetically modified organisms as exotic. Washington and Mississippi have also adopted this definition. Many states indicated that they are confident GEOs would naturally fall under their existing wildlife laws. Overtly specifying this definition, however, could be quite important in keeping transgenic organisms within a state's regulatory framework. For example, if a native fish is taken out of a state's waters, injected with a growth hormone originating from another native fish, then reintroduced to the same state's waters, it would appear that it may still be considered a native fish. If so, it is possible that the fish would not fall under the state's legislation related to exotic species introduction.

The CBP's Exotic Species Workgroup has clearly defined these two issues in the "Chesapeake Bay Policy for the Introduction of Non-Indigenous Aquatic Species." It has defined GEOs as exotics and has detailed a review process that must be followed before an introduction permit is granted. Recall that the CBP was developed under the Special Area Management Planning portion of the federal CZMA. Its purpose is to coordinate state policies that relate to the entire Chesapeake basin from a regional perspective. There was a threefold impetus for the development of this policy: (1) concern over legalization of triploid grass carp in Pennsylvania; (2) crustacean introductions in the James River; and (3) general concerns about zebra mussels (Terlizzi, 1995). The policy explicitly defines GEOs as nonindigenous species, stating that "hatchery-produced hybrids and genetically engineered organisms are also defined as nonindigenous species, even if the parent species or source organisms are indigenous or naturalized."

The policy paper begins, "It shall be the policy of the Jurisdictions in the Chesapeake Bay basin to oppose the first-time introduction of any nonindigenous aquatic species into the unconfined waters of the Chesapeake Bay and its tributaries for any reason unless environmental and economic evaluations are conducted and reviewed in order to ensure that risks associated with the first-time introduction are acceptably low." Under this policy, introductions of non-native species are not totally ruled out. Proposals must first be considered by the respective state marine authorities. If they find the proposal acceptable, it is passed on to the CBP Living Resources Subcommittee, where an ad hoc scientific and technical panel is assembled to examine the proposal and make recommendations to the subcommittee and state. Ultimately, it is up to the state to act on these recommendations and make the decision to approve or deny the proposal for introduction. There is one question remaining with the CBP policy: if faced with a proposal for the introduction of a transgenic marine organism, the Scientific and Technical Advisory Panel must decide what specific risk assessment methodology, if any, it would use to make its determination.

The CBP policy on nonindigenous species introductions is unique; no other states have joined forces in such a comprehensive manner. Although there are regional commissions that coastal states look to for guidance on various issues, none of them has tackled the issue of nonindigenous species introductions in general, or transgenic introductions in particular, as has the CBP.

While we were collecting information from coastal states, it became apparent that many states look to the American Fisheries Society (AFS) for guidance, particularly concerning marine stocking. In 1994, the AFS held a workshop on the "Uses and Effects of Cultured Fishes in Aquatic Ecosystems." Because stocking natural areas with cultured fish is an on-going management procedure in many states, participants in this workshop saw the need to develop a set of guidelines on fish stocking as part of a proactive approach to the conservation of wild stock genes. Workshop participants drafted guidelines for fisheries management teams requiring them to consider the effects of introduced fish on the environment, on native and naturalized biota, and on humans. Twenty-two points to consider were developed under these categories, along with five specific recommendations for stocking in such a way as to reduce the potential for adverse ecological effects. The issues of economic evaluations, public involvement, interagency cooperation, and administrative considerations are also addressed. AFS also has issued a position statement on GEOs, wherein it advocates caution in the use of transgenic fish, improvement in definition and oversight under the Coordinated Framework, and support for research to provide data for rational policy making (Kapuscinski and Hallerman, 1990).

3.4 Marine Aquaculture

The majority of the laws on aquaculture are found at the state level, and they often involve many layers of permit requirements (NRC, 1992). For example, in Maryland before any aquaculture system is started all or any combination of the following may be required: (1) water quality certification; (2) state discharge permit; (3) water appropriations and use permits; (4) federal consistency determination (Coastal Zone Consistency); (5) surface mine permit; (6) aquaculture permit; (7) tidal wetland permit;

(8) critical area approval authority; (9) state non-tidal wetlands and waterways authority; and (10) regional, general, nationwide, and individual permits issued through COE (see Appendix). Several of these permits are related to federal environmental laws, as discussed earlier; however, there are also permits required by Maryland itself through environmental, natural resource, and wetland laws. Information received from other states indicates that multilayered permitting processes, similar to Maryland's, are the norm for aquaculture.

Coastal states appear to be at different levels in grappling with marine aquaculture. Maine is one of the most advanced states for mariculture with what has been termed "one-stop-shopping" for the appropriate permits. The majority of other states with mariculture regulations—including Connecticut, Georgia, Louisiana, and Maryland—have policies specific to shellfish and mollusks. These organisms are usually cultured in open estuarine or nearshore facilities (NRC, 1992). Many states already have experience in leasing usage rights in these areas, so starting aquaculture here is probably easier than in the open ocean. Our coastal state survey also revealed that two states—Massachusetts and Florida—currently are drafting plans for marine aquaculture while six states—California, Delaware, Georgia, Louisiana, Texas, and Washington—simply apply their freshwater aquaculture regulations to the marine environment. Some states—Connecticut, Georgia, Louisiana, Maryland, and Virginia—have adopted specific regulations related to experimental mariculture, and/or to particular organism types, either in conjunction with or separate from freshwater aquaculture regulations.

Unlike freshwater aquaculturists, marine aquaculturists must obtain a lease to use space in the common property resource. This would also be the case for any scientists seeking to secure ocean space to set up an experimental GEO release facility (e.g. floating pens). California, North Carolina, South Carolina, and Maine have all established protocols for granting permits to lease public trust lands. On the other hand, Massachusetts and Florida are still determining whether or not they even have the legislative authority to lease space in the ocean. Therefore, the ease with which an experimental introduction site can be secured will depend upon what state has jurisdiction over the intended waters.

4. CONCLUSION

Because the marine environment is a common property resource, it must be managed in the best interest of current and future generations. As related to our inquiry, this is to say that the federal and state governments must make all efforts to ensure that the risks of any transgenic organism introductions into the marine environment are assessed and found to be acceptably low. In order for a risk assessment to be carried out, the necessary laws and regulations must be in place to trigger oversight that will require such an assessment. Based on the laws presented in this chapter, this is how the process would work for the following situations:

(1) Transgenic salmon used in open-ocean aquaculture: Open ocean is considered federal waters beyond the three-mile state jurisdiction limit. First, the Army Corp of Engineers must be approached to obtain a permit to build or deploy a structure in navigable waters. The FWCA would then require the COE to consult with the appropriate federal agency in charge of marine resources, NMFS. It would then be up to NMFS to assess the ecological implications associated with the structure and with the organism being introduced. Also, because the salmon would be used as a human food source, FDA authorization must be sought and the CVM of FDA would be required to make a determination of ecological risk. What degree of coordination would take place between FDA and NMFS is unknown; however, in this case, existing laws are adequate to trigger a federal examination of the introduction.

However, if the organism is not intended for human consumption, *and* it is not going to be placed in a structure or holding facility in navigable waters, rather released directly into open waters, then there would be absolutely no oversight. Based on current research in transgenic organisms, however, there is no reason to believe that this scenario will take place. It is the only true gap in federal oversight that was encountered in this examination of laws and regulations.

(2) Transgenic oysters placed in an estuary (state waters) to improve water filtration (hypothetical): In this hypothetical case, the oysters will be attached to an existing substrate; no structure will be built requiring a federal permit from COE. The oysters are not intended for human consumption, only for environmental remediation; therefore, an FDA review is not required either. The federal Lacey Act would apply, however, because it provides that no introduction can take place without authorization from the jurisdictional state. It is up to each state to empower

their agencies to fulfill this mandate. Based on the research carried out for this project, it appears that most states have some laws on this subject (14 out of 16 that responded to the questionnaire), although many are based only on lists of prohibited species and GEOs may not be covered. Therefore, proper oversight depends completely on the preparedness of each state for such situations. In this case, no federal involvement would be required, even though the introduction is taking place in the marine environment, with possibly no natural barriers to prevent the wild organism from spreading into the water of neighboring states. In this case, oversight on the introduction is not adequate. Introductions into the marine environment are a regional issue and either require interstate coordination and decision making, or federal oversight.

The most critical point is that the federal agencies and state agencies reviewing the individual introduction cases complete risk assessments. The agencies should develop their own policies on this subject; in fact, it was brought to my attention that NOAA is currently in the process of gathering information for this very topic. In addition to creating policies for internal guidance, these policies could serve as guiding tools for states. Current laws and official policies do not clearly or comprehensively address the introduction of transgenic organisms into the open marine environment; however, there do appear to be adequate triggering mechanisms for oversight to cover most cases.

Means to clarify or better coordinate regulations pertaining to the introduction of transgenic organisms into the open marine environment could include (1) new marine aquaculture legislation; (2) Congressional action to implement ANS Task Force intentional introduction recommendations; (3) advisories by regional fisheries commissions and councils; and (4) actions taken by the coastal states, either individually or collectively. Interest in this issue is apparent among policy makers and regulators at both the federal and state levels indicating that creation of guiding policies and regulations for introducing transgenic organisms may not be too far over the horizon.

APPENDIX

1. Framework of Wildlife and Environmental Laws Related to Transgenic Organism Introductions into the Marine Environment
    CZMA = Coastal Zone Management Act
    MPRSA = Marine Protection, Research and Sanctuaries Act
    NEPA = National Environmental Policy Act

2. Contact List for State Information

California - Robson Collins, Department of Fish and Game, Marine Resources Division
Connecticut - John H. Volk, Department of Agriculture
Delaware - Jim Popham, Department of Agriculture
Florida - Alan Huff and Mark Berigan, Department of Environmental Protection
Georgia - Mike Spencer, Department of Natural Resources, Wildlife Resources Division, Fisheries
Louisiana - Greg Lutz, Louisiana State University, Agricultural Center, Louisiana Cooperative Extension Service
Maine - Ken Honey, Department of Marine Resources
Maryland - Bradley Powers, Department of Agriculture
Massachusetts - W. Blanchard, Department of Agriculture
Mississippi - Gene Robertson, Department of Agriculture and Commerce and Cornell Ladner, Department of Marine Resources
New York - Debra Barnes, NY State Department of Environmental Conservation
Rhode Island - Arthur Ganz, Coastal Fisheries Laboratory
South Carolina - Richard DeVoe, Sea Grant Consortium
Texas - Jim Jones, Department of Agriculture and Bill Harvey, Texas Parks and Wildlife Department
Virginia - T. Robins Buck, Department of Agriculture and Consumer Services
Washington - Candace Jacobs, Department of Agriculture

3. Various Permits needed for Aquaculture (Maryland Requirements)

Water Quality Certification: Federal Clean Water Act - Section 401. This permit covers activities involving filling or construction in wetlands or waterways, including pond construction.

State Discharge Permit: Maryland Environmental Article (COMAR 26.08.04). Deals with aquaculture facilities discharging wastewater to State waters.

Water Appropriations and Use: Maryland Articles (8-802, 8-806, 8-807b). Covers all agriculture uses, including food processing and appropriating an average of 10,000 gallons of water per day or more.

Federal Consistency Determination (Coastal Zone Consistency): Federal Coastal Zone Management Act of 1972. Covers proposed federal activities affecting the State's Coastal Zone. Activities include direct federal agency actions, federal licenses and permits, and financial assistance to state and local governments.

(Permit application reviewed jointly by the Corps of Engineers).

Surface Mine: Maryland Natural Resource Act 7-6A01, 7-6A31 (COMAR .08.05.10.01-.30) Deals with any excavation in excess of one acre that will result in the offsite transport of minerals (any non-fuel earthen material will be considered a mineral).

Aquaculture: Annotated Code of MD - Natural Resources. Covers Title 08 Natural Resources.

Tidal Wetland: Maryland Title 9, Wetlands and Riparian Rights (COMAR 08.05.05 Tidal Wetlands). Deals with the filling and dredging of open water and vegetated tidal wetlands.

Critical Area Approval Authority: Maryland Natural Resources Article 8-1801 through 8-1816 (COMAR 14.15 et seq and 14.19). Deals with all State and local agency sponsored activities/programs affecting the Critical Area (1000' from the mean high water line of tidal waters or the landward side of tidal wetlands).

State Nontidal Wetlands and Waterways Authority: Maryland 8-1201 - 8-1211 (Nontidal Wetlands), 8-203 and 8-801-814 (Waterway construction) (COMAR .08.05.04-.13, COMAR 08.05.04.01). This permit covers all activities involving construction, reconstruction, repair, or alteration of dam, reservoir, or waterway obstruction or any change of the course, current, or cross-section of a stream or body of water within the State, including any changes to the 100 year frequency floodplain of free-flowing, nontidal waters. Also, excavation, fill, changes to hydrology, grading, alteration of vegetation and drainage of nontidal wetlands or nontidal wetlands buffers statewide.

Regional/general, Nationwide, individual permits issued through the Corps of Engineers: Federal Clean Water Act, Section 404. Permits are required for the discharge of dredge or fill material into the nation's waters and/or wetlands for construction of pond embankments, etc.

4. Glossary of Terms Defined in Various Acts

Aquatic Nuisance Species Program:

Intentional Introduction - "The import or introduction of nonindigenous species into, or transport through, an area or ecosystem where it is not established in open waters for a specific purpose such as fishery management. Even when the purpose of such import or transport is not direct introduction into an open ecosystem (e.g., for aquaculture or display in an aquarium), introduction into open waters as the result of escapement, accidental release, improper disposal (e.g. aquarium dumping), or similar releases is a virtually inevitable consequence of the intentional introduction, not an unintentional introduction." (ANSTF, 1994b)

Nonindigenous Species - "Any species or other viable biological material that enters an ecosystem beyond its historic range, including any such organism transferred from one country into another. Nonindigenous species include both exotics and transplants." (ANSTF, 1994b)

Unintentional Introduction - "Introduction of a nonindigenous species that occurs as a result of activities other than purposeful importation, transportation or introduction, such as by the discharge into open waters of ballast water or water used to transport live fish, mollusks or crustaceans for aquaculture or other purposes. Involved is the often unknowing release of nonindigenous organisms without any specific purpose. The virtual certainty of escapement, accidental release, improper disposal (e.g., aquarium dumping), or similar releases of nonindigenous species not intended for such release is considered the consequence of the original intentional introduction, not an unintentional introduction." (ANSTF, 1994b)

Executive Order 11987:

Exotic Species - "all species of plants and animals not naturally occurring, either presently or historically, in any ecosystem of the United States" (EO 11987, 1977)

Introduction - "the release, escape, or establishment of an exotic species into a natural ecosystem" (EO 11987, 1977)

Native Species - "all species of plants and animals naturally occurring, either presently or historically, in any ecosystem of the United States" (EO 11987, 1977)

Federal and State Regulations Relevant to Uncontained 173
Applications of Genetically Engineered Marine Organisms

Lacey Act Amendments of 1981:
Fish or Wildlife - "any wild animals, whether alive or dead, including without limitation any wild mammal, bird, reptile, amphibian, fish, mollusk, crustacean, arthropod, coelenterate, or other invertebrate, whether or not bred, hatched, or born in captivity, and includes any part, product, egg, or offspring thereof" (Public Law 97-79)
Import - "to land on, bring into, or introduce into, any place subject to the jurisdiction of the United States, whether or not such landing, bringing, or introduction constitutes an importation within the meaning of the customs laws of the United States" (Public Law 97-79)

Office of Technology Assessment:
Biotechnology - "Techniques, including both genetic engineering and traditional methods, used to make products and extract services from living organisms and their components" (OTA, 1993).
Genetically engineered/transgenic - "plants, animals, and microorganisms modified by the insertion of genes using genetic engineering techniques" (OTA, 1993).
Genetic engineering - "recently developed techniques through which genes can be isolated in a laboratory, manipulated, and then inserted stably into another organism. Gene insertion can be accomplished mechanically, chemically, or by using biological vectors such as bacteria or viruses" (OTA, 1993).
Genetically modified organisms - "deliberately modified by the introduction or manipulation of genetic material in their genomes. They included not only organisms modified by genetic engineering, but also those modified by other techniques such as traditional breeding, chemical mutagenesis, and manipulation of sets of chromosomes" (OTA, 1993).

5. Pertinent Roles of Relevant Federal Agencies
U.S. Fish and Wildlife Service (U.S. Department of the Interior) - The FWS regulates the import and export of fish and wildlife internationally and across state borders (Endangered Species Act 16 USC Sections 703-712; Marine Mammals Protection Act 16 USC Sections 1531-1543; Migratory Bird Treaty Act 16 USC Section 3371; Injurious Wildlife Act 16 USC Section 152; and Lacey Act 18 USC Section 42,et seq.), as well as comments on proposed construction projects (Fish and Wildlife Coordination Act 16 USC Section 661 et seq.) (Hightower, 1994). With respect to fisheries the "specific responsibilities of FWS include restoration of depleted nationally significant fishery resources, mitigation of fishery resources impaired by federal water-related development, assistance with management of fishery resources on federal and Indian lands, and maintenance of a federal leadership role in the scientifically based management of national fishery resources" (NRC, 1992). "The role of FWS in marine aquaculture remains to be defined, especially with respect to estuarine and coastal species" (NRC, 1992).

U.S. Army Corps of Engineers - The COE maintains and protects the nation's water resources "through the issuance, or denial, of permits authorizing certain activities involving wetlands, and navigable or other waters of the United States" (Hightower, 1994).

National Marine Fisheries Service (U.S. Department of Commerce, NOAA) - "primarily responsible for the management and protection of marine fish, habitat and certain marine animals (16 USC Section 1361 et seq., as amended). To some extent, the NMFS is the marine counterpart to the U.S. Fish and Wildlife Service in regard to fisheries management and protection" (Hightower, 1994). When federal construction and permitting agencies (e.g., COE) solicits comments from the NMFS, "generally, the NMFS reviews [the] construction project applications for any potential impacts to fish and shellfish species and fisheries habitats located in tidal waters and adjacent wetlands. Such reviews also include upstream dams, dikes and other structures that could significantly alter river flows to tidal habitats" (Hightower, 1994). NMFS responsibilities are primarily set out in several major pieces of legislation: (1) the Magnuson Fisheries Conservation and Management Act for conservation of fisheries resources in the U.S. EEZ; (2) the Marine Mammal Protection Act for monitoring, protection, and management of marine mammal stocks in U.S. waters; (3) ESA for monitoring and protection of marine life considered at risk of extinction; and (4) the Fish and Wildlife Coordination Act and the Federal Power Act which provide concurrent responsibilities with the USFWS for protecting aquatic habitat" (NMFS, 1993). The NMFS aquaculture efforts focus mostly on the management of common-property resources and endangered species (NRC, 1992).

Table 1. State Survey Results

State	Regulating agency 1) Wildlife/ marine 2) Marine aquaculture 3) Biotechnology	Marine aquaculture regulations?	Exotic organism introduction regulations	GEOs defined as exotic species?	Regulations on GEO releases?	Member of regional agreement addressing GEO introductions
CA	1) Dept. of Fish & Game, Marine Resources Div. 2) Dept. of Fish & Game 3) ?	Same as freshwater	Yes, list of prohibited species and policy requiring proposals to introduce exotic species to be submitted for approval	No	No, but could be in near future	No
CT	1) Dept. of Env. Protection 2) Dept. of Ag., Bureau of Aquaculture 3) Dept. of Ag.	Specific to molluscan shellfish; no others	Yes, list of prohibited species	No	No	No

Federal and State Regulations Relevant to Uncontained   175
Applications of Genetically Engineered Marine Organisms

State	Regulating agency 1) Wildlife/marine 2) Marine aquaculture 3) Biotechnology	Marine aquaculture regulations?	Exotic organism introduction regulations	GEOs defined as exotic species?	Regulations on GEO releases?	Member of regional agreement addressing GEO introductions
DE	1) Dept. of Nat. Res. and Env. Control 2) Dept. of Ag. with MOU with DNREC 3) ?	Same as freshwater	Yes, policies and protocol for non-native introductions found in CBP	No in state legis-lation, but Yes through CBP	Yes for triploid grass carp. Also, Yes in CBP	Yes...CBP
FL	1) Dept of Nat Res., Div. of Marine Res. 2) Same. 3) Dept. of Env. Protection, Div. of Marine Res. (marine species), Dept. of Ag. and Consumer Services, Div. of Plant Industries (plants)	No, not specific to marine, but some related marine statutes do exist; new comprehen-sive legislation is currently under con-sideration	Yes, list of prohibited species for importation/possession; marine intro-ductions of nonindigen-ous species prohibited; special licenses issued	No	No	No
GA	1) Dept. of Nat. Res., Wildlife Res. Div.; Fisheries 2) Dept. of Nat. Res., Coastal Res. Div. 3) Various	Same as freshwater with a few additional ones for shellfish leased bottoms	Yes, permits required for release	No	No	No
LA	1) Dept. of Wildlife and Marine Res. 2) Same 3) Various	Yes, some specific to oysters and experimental mariculture sites; fresh-water regs apply too	Yes, list of prohibited species releases of nonindigen-ous species prohibited w/o permit	No	No	No

State	Regulating agency 1) Wildlife/marine 2) Marine aquaculture 3) Biotechnology	Marine aquaculture regulations?	Exotic organism introduction regulations	GEOs defined as exotic species?	Regulations on GEO releases?	Member of regional agreement addressing GEO introductions
ME	1) Dept. of Marine Res. 2) Same 3) ?	Yes	Yes, introductions prohibited without permit	No	No regs, however, there is a state commission on testing and release of transgenic organisms	No
MD	1) Dept. of Nat. Res. Tidewater Admin. 2) Same 3) Various	Yes, for oysters and for experimental sites too; freshwater regulations do not apply	Yes, introductions prohibited without permit and procedure for deciding on permit details in CBP	Not in state legislation, but Yes through CBP	Not in state legislation, but Yes through CBP	Yes…CBP
MS	1) Dept. of Marine Res. (Wildlife under separate dept.) 2) Dept. of Marine Res. and Dept. of Ag. 3) None	Yes, Aquaculture Act of 1988 has specific references to marine environment	Yes…non-native species used in aquaculture require permits	Yes… in Aquaculture Act of 1988, Section 79-22-9	Not for general GEO experiments, but Yes for aquaculture GEOs	No
NY	1) Dept. of Env. Conservation 2) Same 3)?	Yes	Yes, permits required for introduction or release of marine organisms from waters outside state	No	No	No

Federal and State Regulations Relevant to Uncontained
Applications of Genetically Engineered Marine Organisms

State	Regulating agency 1) Wildlife/ marine 2) Marine aquaculture 3) Biotechnology	Marine aquaculture regulations?	Exotic organism introduction regulations	GEOs defined as exotic species?	Regulations on GEO releases?	Member of regional agreement addressing GEO introductions
RI	1) Dept. of Env. Mgt, Div. of Fish & Wildlife 2) Same 3) Same	No, no specific aquaculture regulations for freshwater or marine; determined case by case	Nonindigenous species are not permitted, although there is no specific regulation stating this	No, but may be in near future	No	No
SC	1) Dept. of Nat. Res. 2) Dept. of Nat. Res., Dept. of Health & Env.; Dept. of Ag. 3) Dept. of Health & Env.	No, not specific to marine; considered species by species	Yes, permits required for introduction or release of non-native or non-established wildlife	No	No	No
TX	1) Parks & Wildlife Dept. 2) Dept. of Ag. (facilities); Parks & Wildlife (exotic species) 3) ?	Same as freshwater	Yes, list of prohibited species; permits required for other exotic species; potentially harmful species prohibited	No	No	No
VA	1) Marine Res. Comm. 2) Same 3) ?	Yes, pertaining to striped and hybrid bass	Yes, list of approved species; permits required for all other; also, CBP protocol	Not in state legislation, but Yes through CBP	Not in state legislation but they are currently being worked on; Yes in CBP	Yes...CBP

State	Regulating agency 1) Wildlife/ marine 2) Marine aquaculture 3) Biotechnology	Marine aquaculture regulations?	Exotic organism introduction regulations	GEOs defined as exotic species?	Regulations on GEO releases?	Member of regional agreement addressing GEO introductions
WA	1) Dept. of Fish & Wildlife 2) Same 3) Dept. of Ag.	Same as freshwater	Yes, permit required for release; proposals for introduction must contain complete report as required by state Env. Policy Act	Yes	Not in general, however, insect pests and plant diseases are addressed	No

REFERENCES

Aquatic Nuisance Species Task Force (ANSTF). Fish and Wildlife Service (FWS) and National Oceanic and Atmospheric Administration (NOAA). 1994a. Report to Congress: Findings, Conclusions, and Recommendations of the Intentional Introductions Policy Review. Washington, DC: United States Government 1-53.

Aquatic Nuisance Species Task Force (ANSTF). 1994b. Aquatic Nuisance Species Program. Washington: U.S. Fish and Wildlife Service and National Oceanic and Atmospheric Administration.

Beatley, T., D.J. Brower and A.K. Schwab. 1994. An Introduction to Coastal Zone Management. Washington, DC: Island Press.

Chesapeake Bay Program (CBP). 1993. Chesapeake Bay Policy for the Introduction of Non-Indigenous Aquatic Species. CBP/TRS 112/94. Washington, DC: U.S. Environment Protection Agency 1-29.

Clugston, J.P. 1986. Strategies for reducing risks from introductions of aquatic organisms: the federal perspective. Fisheries 11:26,28-29.

Colligan, M. 1996. Personal communication. (Colligan is with NMFS, Northeast Regional branch, Gloucester, MA.)

Community Nutrition Institute (CNI). 1995. Transgenic Fish: the next threat to marine biodiversity. Washington, DC: Community Nutrition Institute 1-15.

Courtenay, W. R. Jr and J. D. Williams. 1992. Dispersal of exotic species from aquaculture facilities, with emphasis on freshwater fishes. In: A. Rosenfield and R. Mann (eds.). Dispersal of Living Organisms into Aquatic Ecosystems. College Park, MD: Maryland Sea Grant College.

Ewart, J.W., J. Hankins and D. Bullock. 1995. State policies for aquaculture effluents and solid wastes in the Northeast region. NRAC Bulletin #300. North Dartmouth, MA: Northeastern Regional Aquaculture Center, University of Massachusetts, Dartmouth 1-24.

Farmer, S.B. and B.L. Buniva. 1995. Virginia passes biotechnology legislation for economic development. Genetic Engineering News 14:24.

Findley, R.W. and D.A. Farber. 1992. Environmental Law in a Nut Shell. St. Paul: West Publishing Company).

Ganz, A. 1995. Personal communication. (Ganz is with the Coastal Fisheries Laboratory in Wakefield, RI.)

Gladwell, M. 1988. Towns restricting tests of altered organisms. Washington Post March 20 H5.

Guralnik, D.B. (ed.). 1986. Webster's New World Dictionary. Second College Edition. New York: Prentice Hall Press.

Hallerman, E.M. and A.R. Kapuscinski. 1992. Ecological and regulatory uncertainties associated with transgenic fish. In: C.L. Hew and G.L. Fletcher (eds.). Transgenic Fish. New Jersey: World Scientific 209-228.

Hightower, M. 1994. Permits, Licenses, Certificates and Requirements Affecting Aquaculture Regulations. Broyan, TX: Sea Grant Marine Advisory Service 1-60.

Hoffmann, D. 1988. The biotechnology revolution and regulatory evolution. Drake Law Review 38:471-550.

Interagency Task Force on Biotechnology S.O. (ITFB). 1986. Biotechnology: California permits and regulations. California: State of California 1-212.

Johnson, R.B. 1995. U.S. policy for marine aquaculture. Journal of Marine Biotechnology 67-68.

Kapuscinski, A.R. and E.M. Hallerman. 1990. Transgenic fishes. Fisheries 15:2-4.

Kelly, G. 1996. Presentation given at the Open Ocean Aquaculture Conference, May 8-10, Portland, ME.

King, L.R. and F.D. Jennings. 1988. The executive and the oceans: three decades of United States marine policy. Marine Technology Society Journal 22(1):17-23.

Lassuy, D.R. 1994. Aquatic nuisance organisms: setting national policy. Fisheries 19:14-17.

Matheson, J. 1997. Personal communication. (Matheson is with CVM in Rockville, MD.)

McClain, W.E. Jr. 1991. U.S. Environmental Laws. Washington, DC: Bureau of National Affairs.

McCoy, H.D. 2nd. 1993a. Open ocean fish farming: part one. Aquaculture Magazine 19:66-74.

McCoy, H.D. 2nd. 1993b. Open ocean fish farming: part two. Aquaculture Magazine 19:60-67.

Metrick, S. 1995. Personal communication. (Metrick is from the office of U.S. Senator John F. Kerry, Washington, DC.)

National Research Council (NRC). 1992. Marine Aquaculture: Opportunities for Growth, Washington, DC: National Academy Press.

National Biological Impact Assessment Program (NBIAP). 1995. On-line service update of state regulations on biotechnology <news@nbiap.biochem.vt.edu>.

National Marine Fisheries Service (NMFS). 1993. Our Living Oceans: Report on the Status of U.S. Marine Living Resources, 1993. Washington: United States Department of Commerce 1-156.

National Research Council (NRC). 1992. Marine Aquaculture: Opportunities for Growth, Washington: National Academy Press.

Office of Technology Assessment (OTA). 1993. Harmful Non-Indigenous Species in the United States. Washington, DC: U.S. Government Printing Office.

Peoples, R.A. Jr and J. Troxel. 1996. Personal communication. (Peoples and Troxel are with the USFWS, Arlington, VA)

Peoples, R.A., Jr., J.A. McCann and L.B. Starnes. 1992. Introduced organisms: Policies and activities of the U.S. Fish and Wildlife Service. In: A. Rosenfield and R. Mann (eds.). Dispersal of Living Organisms Into Aquatic Ecosystems. College Park, MD: Maryland Sea Grant College 325-352.

Plante, J.M. 1996a. Scallop project blazes open-ocean farming trail. Fish Farming News 20-21.

Plante, J.M. 1996b. U.S. industry breaks barrier: first sea cage anchored successfully in federal waters. Fish Farming News 4(July/August):1,24.

Pollack, S. 1991. Suit over offshore salmon farm plan. National Fisherman 72:8.

Powers, D. 1992. Memorandum of understanding by and between the California Department of Fish and Game and Stanford University regarding genetic engineering of red abalone.

Regal, P.J. 1994. Scientific principles for ecologically based risk assessment of transgenic organisms. Molecular Ecology 3:5-13.

Stevens, L. 1996a. Farm bill passes: aquaculture language axed by Young. Fish Farming News 11:1,17

Stevens, L. 1996b. Aquaculture bill gets one last shot in '96. Fish Farming News 4(July/August):3.

Stevens, L. 1996c. National Aquaculture Development Plan: industry offers substantive ideas on rewrite. Fish Farming News 4(July/August):7.

Terlizzi, D. 1995. Personal communication. (Terlizzi is Chair of the Exotic Species Workgroup of the Chesapeake Bay Program, Annapolis, MD.)

United States Congress. 1981. Lacey Act Amendments of 1981. Public Law 97-79 1-9.

Chapter 6

Economic Analysis of Introduced Genetically Engineered Organisms

DIANE HITE AND JOHN J. GUTRICH
Ohio State University

1. INTRODUCTION

Previous chapters of this book have dealt with issues such as the current regulatory framework for risk assessment of transgenic organisms, biotechnology legislation and related policies, and the ecological risks associated with accidental releases of genetically engineered organisms (GEOs). In this chapter, we consider these issues in an economic context. It has often been noted that the words "ecology" and "economy" derive from the same root, and in fact the fields of study share a number of common characteristics (Costanza et al., 1997a; Costanza, 1991; Jansson, et al., 1994; Krishnan et al., 1995; Milon and Shogren, 1995; Norgaard, 1988; Rapport and Turner, 1977; Rothschild, 1990). Economic and ecological systems are both highly complex and in constantly evolving dynamic equilibrium; shocks to either type of system can cause either temporary or long-term disequilibrium to occur. Here, we utilize the economic approach of cost-benefit analysis to evaluate possible alterations and/or shocks to current economic conditions of natural marine systems that may result from the introduction of transgenic marine organisms.

The ability of scientists to genetically manipulate marine species has the potential to generate great benefits for humans in areas such as aquaculture, marine animal health, biofilm/bioadhesion, and bioremediation (Colwell, 1987; Lundin and Zilinskas, 1995; Zilinskas et al., 1995). For example, in an attempt to foster increased yields in intensive aquaculture operations, scientists can create transgenic finfish and shellfish with enhanced growth rates and increased ability to survive adverse environmental conditions (Chen et al., 1996; Devlin et al., 1994; Shears et al., 1991). However, as discussed in Chapter 3, utilization of transgenic organisms in or near natural marine ecosystems poses a potential ecological risk in case of accidental release. Introduction of a transgenic marine organism into a habitat, in which it utilizes novel resources unavailable to the parent species, may alter the natural composition, abundance, and diversity of organisms and communities of organisms within that environment. In turn, these alterations may adversely affect the ecosystem structure and function (Cairns and Pratt, 1986; Tiedje et al., 1989).

Benefits, costs, and risks are translated into economic terms when the value that humans derive from the various components of the environment (i.e., the economic value of ecosystem services) is affected. In this chapter, our main intent is to (1) provide the conceptual framework for a cost-benefit analysis of introduced marine GEOs; (2) identify the potential costs and benefits of introduced GEOs from a neoclassical, environmental, and ecological economic approach; (3) indicate costs to industry likely to be incurred when implementing novel risk management protocols; and (4) apply an environmental economic analysis to determine an efficient level of expenditure on containment.

2. COST-BENEFIT ANALYSIS OF INTRODUCED MARINE GEOS

In order to assess the economic impacts of introduced marine GEOs, we provide the framework for a cost-benefit analysis (CBA) and indicate some key potential costs and benefits. The objective of CBA is straightforward: to assess the advantages and disadvantages of a proposed public policy (Dasgupta and Pearce, 1978). CBA is an economic tool taking a societal standpoint and is often utilized for programs that have unmarketed (or unpriced) outputs such as alterations in environmental quality and ecosystem function (Field, 1994). The main principle is that CBA is a test to decide whether the proposed program or policy could make at least one person better off and no one worse off, after compensation (Randall, 1987). In the current context, a CBA would attempt to determine whether the proposed release of marine GEOs would generate sufficient benefits to increase overall social welfare after compensating any and all losers. CBA represents just one component of the decision-making process and serves as a complement to the political process, not a substitute for it (Hitzhusen, 1984).

Quantification of all societal economic costs and benefits that could result from the introduction of marine GEOs will be a complex and challenging task requiring interdisciplinary collaboration. For example, in order to estimate economic effects from alterations of the environment, information must be provided on the probable ecological effects caused by the release of a marine GEO on a case-by-case basis. Thus, ecologists will need to collaborate with economists in the risk assessment process in order to establish monetary estimates of any positive or negative impacts on marine ecosystems. Here, we provide a framework for assessing the economic effects of introduced marine GEOs by addressing potential market effects, external effects, empirical methods for estimating benefits and costs, and regulatory and compliance costs.

2.1 Potential Market Effects

To measure the full societal costs and benefits of marine GEOs, one can initially conduct analyses from the conventional neoclassical economic approach. Neoclassical economics is generally concerned with outcomes for goods that can be produced and bought in an organized market (e.g., price, quantity, allocation, distribution, etc.). A market is an institution in which buyers and sellers carry out mutually agreed upon exchanges (Field, 1994). If goods or services are traded in the market, there are well established and accepted empirical techniques for measuring social welfare changes (Lipton et al., 1995). A standard methodology for measuring the societal benefit of resources traded in a market is the estimation of producer and consumer surplus utilizing market price and quantity data (Tietenberg, 1996). In order to understand the potential gains in the market from introduced marine GEOs, we review some key concepts of neoclassical economic benefit estimation.

2.1.1 Consumer surplus: a measure of benefits to consumers

Marine GEOs provide obvious benefits from the standpoint that they may provide a means to increase food supplies to the human population. It is a generally recognized phenomenon in economics that as supplies of a given commodity increase, prices decrease, thus making the commodity available to more individuals. The relationship that traces out the quantities demanded of a commodity at various price levels is the demand curve (see Figure 1). The demand curve for a commodity starts with zero units purchased at high price levels, and as prices decrease, the quantity of the commodity demanded increases.

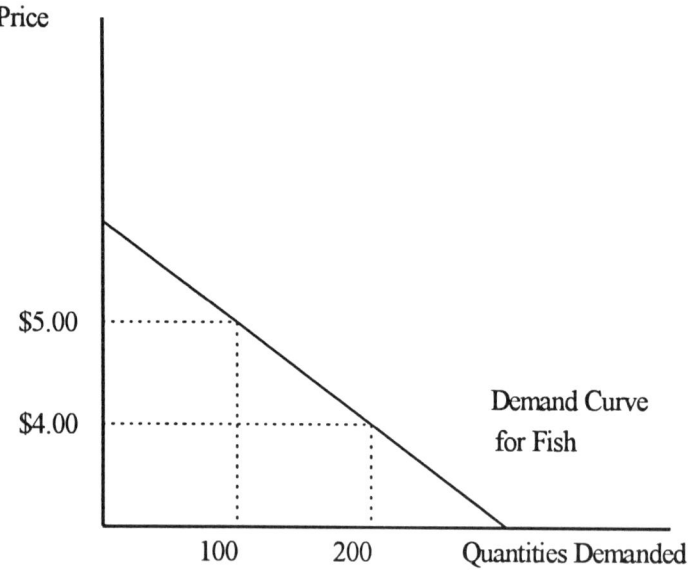

Figure 1. Demand for Marine Fish

Given a market price, the consumer decides the quantity to purchase by choosing the amount which maximizes her or his individual net benefit (Tietenberg, 1996). In consideration of the hypothetical scenario for marine fish illustrated in Figure 1, at a price of $5.00 per unit of fish, consumers demand only 100 units. However, once the price of fish decreases to $4.00 per unit, an additional 100 units will be demanded. When the price is $4.00, those willing to pay more than $5.00 per unit of fish are receiving two types of benefit: (1) the benefit from consumption of the fish, and (2) the benefit given by the difference between the price they are willing to pay and the actual market price per unit of fish. The fact that individuals exist in the market who are willing to buy fish at higher prices than they actually face in the market leads to a measure of economic benefit to consumers called consumer surplus (CS). Thus, consumer surplus is the consumers' net

Economic Analysis of Introduced Genetically Engineered Organisms

benefit, represented by the area under the demand curve minus the area representing the cost to the consumer (see Figure 3).

2.1.2 Producer surplus: a measure of benefits to producers

An analogous measure of economic benefit can be developed for producers of fish, whether they are producers of cultured fish (i.e., aquaculturists) or wild fish (i.e., fishermen). The relationship between the price of fish and the quantities of fish produced is called the supply curve (see Figure 2). The supply curve starts at zero quantities produced at low prices; then quantities supplied increase as prices increase. If prices are at $5.00 per unit, then producers who are willing to supply fish at lower prices obtain the benefit not only of the prevailing price, but also the difference between the market price and the actual price they are willing to accept for the fish. This measure of economic benefit is known as producer surplus (PS). Thus, the producers' net benefit is the entire area above the supply curve up to the price line (Figure 3).

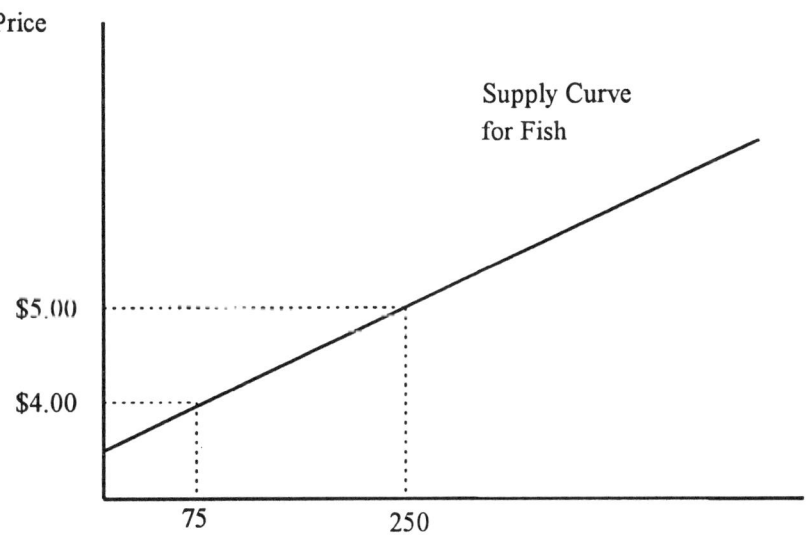

Figure 2. Supply of Marine Fish

2.1.3 Total social surplus: a measure of social market benefits

To obtain the total net economic benefit for fish traded in a market, it is necessary to examine the behavior of consumers and producers simultaneously. In economic terms, this benefit is called total social surplus (TSS), and consists of the sum of PS and CS in equilibrium. While not an exact measure of social welfare, the sum of consumer and producer surplus provides a useful approximation of the net benefit of a good or service (Lipton et al., 1995). Equilibrium occurs at the point where supply of fish equals the demand for fish; the price and quantity at this point are called the equilibrium price (Pe) and equilibrium quantity (Qe). Thus, in the market equilibrium illustrated in Figure 3, TSS is given by the total area under the demand curve and above the supply curve for fish up to the point where the curves intersect (i.e., TSS = CS+PS). At the equilibrium price and quantity, net benefits are maximized, as consumers and producers maximize their respective surpluses, and the market clears.

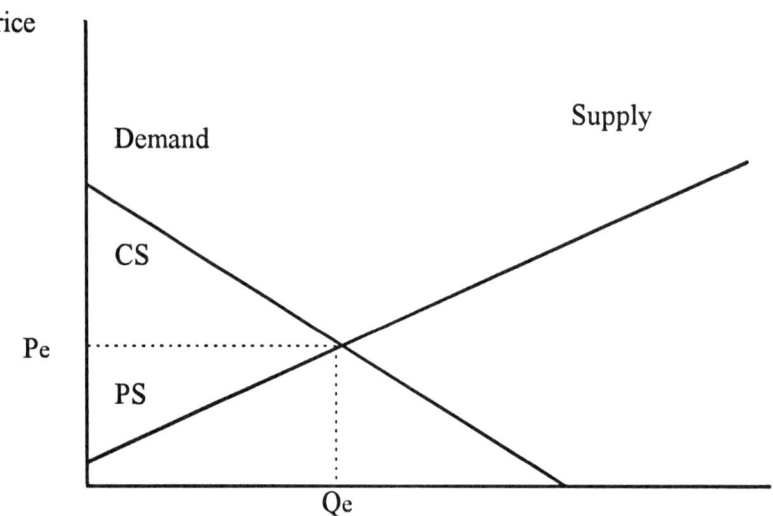

Figure 3. Equilibrium in the Market for Fish

Economic Analysis of Introduced Genetically Engineered 187
Organisms

2.1.4 Potential market benefits from introduced marine GEOs

The economic benefits generated by the introduction of a marine GEO can be illustrated utilizing neoclassical analysis to examine the example of salmon, a commonly cultured and genetically altered marine fish. Suppose that the equilibrium illustrated in Figure 3 represents supply and demand in the market for wild salmon only. It should be obvious that any force that would increase the supply of fish and lower the equilibrium price would mean an increase in net market benefits. Such a supply increase could be obtained by the introduction of genetically engineered salmon. The increase could become manifest in two ways: (1) an increase in the number of aquacultured transgenic salmon brought to market or (2) an increase in the size of aquacultured transgenic salmon (i.e. weight per fish) offered in the market. For example, A/F Protein Inc. of Massachusetts is currently licensing, in New Zealand and Great Britain, transgenic salmon that reach maturity in two years instead of the normal three years (MacKensie, 1996). Higher maturity rates result in lower economic costs to fish farmers by reducing the risk of disease or escape of aquacultured fish. Such transgenic salmon create an opportunity to bring more salmon to market and thus increase supply. Further, the European Seafood Buyer has reported size enhancement of transgenic Pacific salmon to a mass 11 times that of the normal parent (Infofish, 1995). Thus, increased body size of transgenic marine fish may also provide the opportunity to bring larger salmon to market.

Figure 4 (below) illustrates the increase in net benefits (i.e., TSS) that could occur from the introduction of such engineered salmon. Before the introduction of the GEO, supply of salmon is S, price is P, and quantity demanded is Q; CS is given by area a, and PS is given by area b+c, so that TSS is equal to a+b+c. The effect of the new supply source is to shift the supply curve of salmon from S to S'. This shift results in a price reduction from P to P', quantities demanded increase from Q to Q', and TSS increases by d+e+f+g. Thus, the introduction of GEOs can clearly produce major economic benefits via decreased prices and increased consumption.

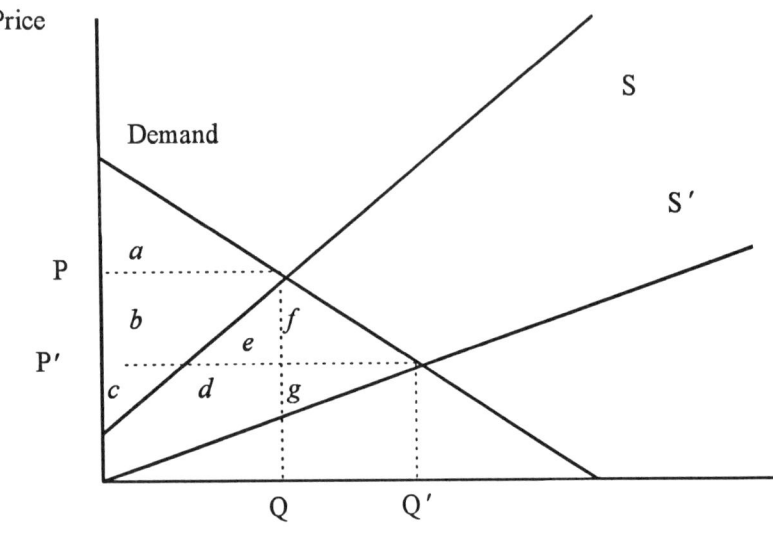

Figure 4. Increased Economic Benefits from Introduction of GEOs

2.2 Potential External Effects

The foregoing analysis of social benefits takes into consideration only one part of the analysis of costs and benefits of GEO introductions; the benefits from consumption and production. However, the "external effects" marine GEOs may have on the world's fisheries, as well as the effects that regulations may have on economic outcomes, require an extension of the neoclassical approach. Thus, we broaden our analysis by applying environmental and ecological economics.

Broadly defined, environmental economics is the study of environmental problems utilizing the analytical ideas of economics (Field, 1994). Beyond the valuation of market goods, environmental economics includes the study of external effects (i.e., externalities) associated with the production and/or consumption of market goods and services. Externalities are unpriced by-products of market goods, sometimes positive (e.g., scenic views created), but often recognized as negative (air pollution, sediment deposition, degradation of biodiversity, eutrophication from nutrient runoff, noise pollution, etc.). Negative externalities occur when the actions of one economic agent (e.g., firm or household) affect the welfare of another, who is not compensated for ensuing damages (Barbier, 1989).

The concept of an externality is pertinent for our discussion because marine GEOs may generate external benefits or costs to society that are not transacted in a market. If the GEOs are causing damages that are not accounted for in the supply curve, then costs will be underestimated and the equilibrium price and quantity will not maximize net social benefits. In Figure 5, the market supply curve represents the private marginal cost of producing and utilizing GEOs, while the full costs to society are represented by the marginal social cost curve. If the full social costs are not considered, the quantity of GEOs will be oversupplied. Similarly, underestimating benefits would result in a demand curve which when intersecting with the supply curve would also not produce equilibrium conditions that maximize net benefits. Thus, a CBA of introduced marine GEOs must attempt to internalize externalities in order to assess full social costs and benefits in order to maximize net benefits to society at large. Here, we address some pertinent potential positive and negative externalities associated with the introduction of marine GEOs.

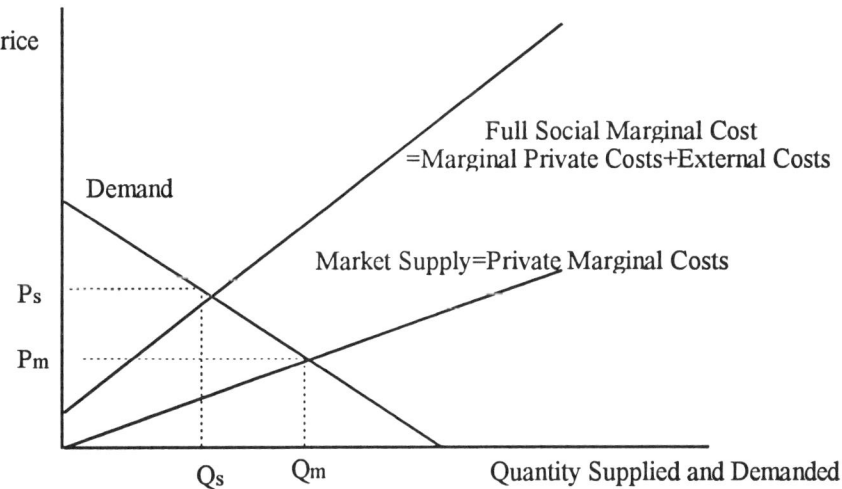

Figure 5. External Effects of Introduced Marine GEOs

2.2.1 Potential positive external effects of marine GEOs

One potential positive externality of marine GEO introductions involves substitution of aquacultured transgenic fish for natural fish in the market. By substituting GEOs for the parent species as a food source, pressure on marine fisheries may be reduced, possibly saving some species from a population collapse (Kaufman and Dayton, 1997; Powers, 1995). The world's natural fisheries are under stress as overfishing occurs in an effort to meet growing global demand for seafood (Botsford et al., 1997; FAO, 1995a; Weber, 1995). Worldwide seafood demand is growing rapidly and is expected to increase 70% in the next 35 years (JSA, 1992; NSTC, 1995). Yet, the Food and Agricultural Organization (FAO) of the United Nations (1995b) reports that in 6 of 11 major Atlantic and Pacific fishing regions, more than 60 percent of all commercial fish stocks either are depleted or are being fished to their biological limit. In turn, closing of fisheries can lead to high local unemployment as manifested in eastern Canada, where 40,000 jobs were eliminated when the cod fishery closed in 1992 (Clayton, 1995; Garcia and Newton, 1994; WRI, 1996). Economic opportunities from employment in the fishing industry either directly or indirectly support some 200 million people worldwide (WRI, 1996). The contained culturing of GEOs may help maintain employment levels in the fishing industry worldwide by allowing natural stocks to remain at higher levels due to reduced harvest pressure.

Aquaculture has already begun to meet growing demand for seafood as natural global catches decline. Although, marine natural harvest has declined from 86 million metric tons in 1989 to 84 million metric tons in 1993, total global fish harvest has continued to climb because of increasing aquaculture production (FAO, 1995b, 1993). According to the FAO, aquaculture production has risen from 7.4 million tons in 1984 to 20.5 million tons in 1994, supplying almost 20% of the seafood eaten in 1994 (Christensen, 1997). Figures 6 and 7 represent the growth of salmon aquaculture in Canada, Japan, the United Kingdom, and the United States between 1985 and 1994. Canada displays the most significant change with current production approaching 65,000 metric tons annually. Aquacultured salmon has increased from only 0.5% of total Canadian salmon production in 1984 to 41.4% in 1994 (FAO, 1996). Aquaculture production is expected to increase to over 22 million metric tons by the year 2000, representing from 25% to 33% of the world's harvest (Norse, 1993; Parker, 1992)

Economic Analysis of Introduced Genetically Engineered Organisms

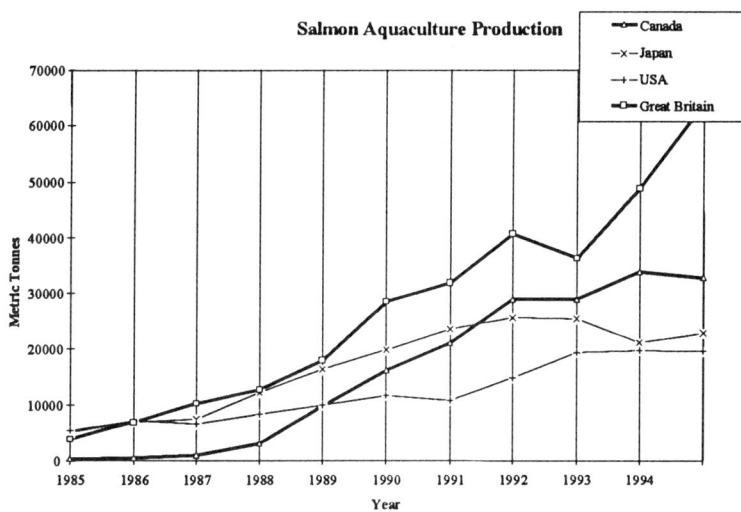

Figure 6. Aquacultured Salmon Production, 1985-1994

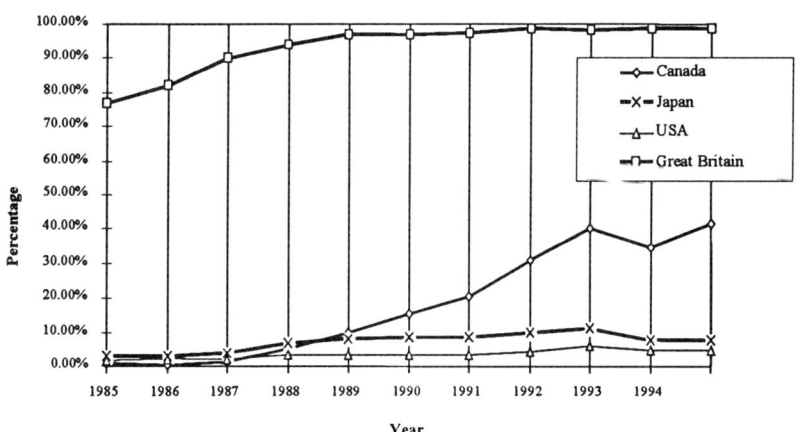

Figure 7. Aquacultured Salmon as Percent of Total Salmon Production, 1985-1994

In economic terms, global aquaculture sales reached $22 billion in 1990 and grew to $33.5 billion by 1994 (FAO, 1996; Fischetti, 1991). In the United States alone, aquaculture sales have grown from $464 million in 1984 to $795 million in 1994 (estimates in 1996 dollars, according to NMFS; Christensen, 1997). High demand in the market with limited supply is expected to continue to drive up prices and to make fish farming more lucrative in upcoming years (Parker, 1992). Thus, aquaculture provides the mechanism by which introduction of marine GEOs may provide large potential economic gains by meeting growing seafood demand and ensuring continued benefits to natural fisheries by protecting wild stocks.

Further potential external benefits from introduced marine GEOs could be derived from positive ecological effects that restore goods and maintain marine ecosystem services. Transgenic marine bacteria utilized for bioremediation may have major positive ecological economic impacts (Frederick and Egan, 1994). From the neoclassical perspective, assessment of economic impacts would be straightforward. Bioremediation of oil spills by introduced GEOs may present significant cost savings to private firms and society from the restoration of the condition of goods and services provided by the marine environment (e.g., fisheries, recreational fishing locations, etc.) to pre-spill levels. Transgenic organisms that could increase the rate of recovery of valuable fish populations (or other market goods and services) at a lower cost than current bioremediation practices would be beneficial.

Transgenic organisms utilized in bioremediation may also provide a means to clean oil spills more thoroughly than previously possible, thereby restoring populations and communities of marine species and functional levels of ecosystem services. Economic valuation of the benefits derived from restoring non-market marine goods and services will require valuation of externalities and ecosystem services. However, Pearce (1985) has argued that a disparity exists between analyses of externalities and ecological economic effects, stating that there is nothing in the conventional concept of an external cost to account for the decay of ecological processes themselves. Thus, it is essential that a CBA of introduced marine GEOs analyze such "ecological economic" effects.

Ecological economics differs from conventional and environmental economics in terms of the breadth of perception of environmental problems and the importance attached to environment-economy interactions (Costanza et al., 1997a; Costanza, 1991). Ecological economics attempts to estimate values of ecosystem goods and services (e.g., wetland sewage treatment, global climate control, etc.) that are often long-term in nature, generally not traded in markets, and for which information about their contribution to human well being is poor (Costanza, 1991).

Ecological services are ecosystem functions that are perceived to support and protect human activities (Barbier et al., 1994). Ecosystem services maintain biodiversity and the production of ecosystem goods such as seafood, biomass fuels, natural fiber, and pharmaceuticals (Daily, 1997; Gutrich, 1997). For example, a fish in the sea is a product of highly complex interactions with other ecological sectors such as prey species, phytoplankton that release vital oxygen, and microorganisms that maintain a flow of essential nutrients or inhibit destructive eutrophication through cycling. Peterson and Lubchenco (1997) identified pertinent services of marine ecosystems including transformation, detoxification, and sequestration of pollutants and societal wastes, as well as global materials cycling (e.g., CO_2 utilization).

Ecosystem services are seldom reflected in resource prices or considered in economic indicators utilized by developed countries (Dixon and Hufschmidt, 1986; Daly, 1991; Ehrlich and Ehrlich, 1992). Thus, enhanced rehabilitation from introduced marine GEOs may foster restoration of ecosystem function and structure to a level at which it continues to provide services that are not easily valued within neoclassical economics. Extending the CBA into ecological economics through collaboration between ecologists and economists in the risk assessment process is an essential component of any attempt to capture positive or negative ecological economic effects that may result from alteration of ecosystem services.

2.2.2 Potential negative external effects of marine GEOs

Perhaps the most significant economic costs that can arise from the introduction of a marine GEO would result from release of an organism that subsequently became established as an "exotic" species (see Chapter 3). First, one can imagine a scenario in which introduced GEOs deplete the abundance and diversity of a community of economically valuable marine fish by outcompeting the natural stocks. Depletion of natural fish stocks from introduced GEOs can lead to substantial economic costs. Introduced GEOs may parallel the effects from introduced exotic species such as the ctenophore (a "comb jelly" organism) which was accidentally introduced into the Black and Azov Seas. The exotic ctenophore voraciously fed on native fish eggs leading to the collapse of local fisheries, causing an estimated $250 million in damage (Travis, 1993).

Marine fish account for roughly 20% of protein consumed by humans worldwide and certain developing countries display levels of consumption as high as 60% (Gutrich, 1997; Kaufman and Dayton, 1997; Norse, 1993). An estimated 950 million people depend on fish as their primary source of protein (WRI, 1996). In turn, the global value of marine fisheries is substantial, estimated from first-sale fishery revenues to be between $50 and $100 billion (Kaufman and Dayton, 1997; Botsford at al., 1997). Thus, depletion of natural populations of foodfish by introduced GEOs could jeopardize highly valuable goods that are essential in the human diet.

Second, there is the possibility that introduced GEOs could drive out marine species that are highly valued for recreational activities. In 1985, the U.S. Fish and Wildlife Service (FWS) estimated the value of the U.S. sport fishing industry at $28.2 billion; this figure is forecasted to double by the year 2030 (Parker, 1992; USFWS, 1988). In 1990, gross expenditures for recreational fishing in the U.S. had already reached $38.8 billion (USFWS, 1993). Habitat perturbation could result in effects that could extend to aquatic avian species or mammalian species that are also valued for hunting and non-consumptive uses (e.g., photography). In 1991, U.S. gross hunting expenditures were an estimated $12.3 billion (USFWS, 1993).

Many Caribbean islands generate large revenues from non-consumptive ecotourist attractions that depend upon the maintenance of natural species such as reef fishes, corals, and even associated invertebrates, viewed by snorkeling, diving, and with glass bottomed boats (Peterson and Lubchenco, 1997). For example, total gross revenues were estimated at $23.2 million for dive-based tourism at the Bonaire Marine Park, a marine sanctuary in waters off the Caribbean island of Bonaire (Dixon et al., 1994; Dixon et al., 1993; Scura and van't Hof, 1993). GEOs introduced into aquaculture facilities to bolster local island economies may escape and subsequently deplete the abundance and diversity of recreational fish populations causing substantial local economic losses from declining tourist revenues. In the case of Bonaire, Dixon and associates (1994) indicated that "since there are few other attractions on the island, a decrease in the level of protection and degradation of the marine resource would result in both loss of ecological and economic benefits; any loss of reef and water quality and reduction in the fish population would result in divers shifting their demands to other islands." Thus, in considering localized economies, it is important to assess possible economic losses from introduced GEOs that may be incurred in economic sectors for which there are no domestic substitutes.

Introduced GEOs may also increase costs for current production activities in other areas of the economy. For example, it is possible that introduced GEOs could act as fouling agents similar to one of the most notable introduced aquatic, exotic species of the decade, the zebra mussel. Upon accidental introduction into the Great Lakes, the zebra mussel, *Dreissena polymorpha*, reproduced rapidly and established colonies on a variety of substances, including stone, steel, concrete, wood, plastic, and glass. Zebra mussels have interrupted production activities by fouling water intake pipes at public water supply facilities, electric power plants, and conventional and nuclear fuel power plants (Zilinskas et al., 1995). Further, zebra mussels have damaged cooling systems of ships, harbor structures, canal locks, and flood control mechanisms. Total economic damage from the zebra mussel is substantial, estimated to reach $3 to $5 billion by the year 2000 (ANSTF, 1992; Cohen and Carlton, 1995). Thus, beyond economic costs to natural harvest and recreational fishing, introduced GEOs (acting as "exotics") may cause significant direct economic costs to various economic sectors related to the marine environment, warranting consideration in a CBA.

A number of the economic, ecological, and environmental costs and benefits associated with aquaculture are relatively well known. However, certain costs and benefits from introduced GEOs may be incremental to those related to fish farming and other marine biotech operations. To the extent that GEOs will be utilized in aquaculture production, damages to the benthic layer of the ocean may be exacerbated. The ocean floor in the vicinity of conventional aquaculture facilities has displayed signs of deterioration due to the wastes that are a byproduct of production (Kaufman and Dayton, 1997). Fish waste creates anoxic conditions (i.e., low oxygen) and degrades seafloor biota beneath offshore aquaculture cages, decreasing the abundance and diversity of micro- and macroorganisms (Morton, 1996). The degradation of the ocean floor has uncertain effects on the overall health of the nearby ecosystems, and may contribute to conditions that are detrimental to natural fisheries. Introduced GEOs with enhanced body size may release greater quantities of waste products and thereby increase economic costs related to degradation of the local benthic community.

Another potential negative external effect from introduced GEOs is overfishing. Increased fishing effort can lead to the external effect of overfishing, and introduced GEOs may indirectly foster increased effort.

For example, if introduced GEOs increased the supply of fish on the market, prices would decrease. Lower prices per unit of fish may provide an incentive to commercial fishing enterprises to increase their fishing effort, resulting in overfishing of valuable species.

Such a counterproductive outcome may be a realistic possibility as evidenced by trends in fishing effort and total global catch. From 1950 to 1989, marine natural harvest increased nearly fivefold, peaking at 86 million metric tons in 1989 before dropping in subsequent years (FAO, 1995c, 1993; WRI, 1996). However, fishing effort has also dramatically increased over time. Powers (1995) shows that the total number of fishing vessels worldwide has increased from 2.2 million vessels in 1970 to roughly 3.3 million by 1989. FAO also indicates that over the same period the number of boats classified as large doubled from 585,000 to 1.2 million (Weber, 1995). Due to the combination of these two factors, total tonnage of fishing vessels has grown from 13.6 to 25.9 million metric tons, an increase of 91% (FAO, 1995b). In an attempt to recoup private gains lost to falling prices or to bolster profits in times of high demand, fishing effort can increase to the detriment of natural stocks.

Such a scenario unfolded recently in the Pacific salmon fishery along the U.S. and Canadian coastline. Even though scientists in both countries warned that two salmon species, chinook and coho, are in danger of collapse from overfishing, the U.S. and Canada failed to reach an agreement for the 1997 Pacific salmon fishing season (Shanks, 1997). However, historically low salmon prices are expected to result in increased catch effort that will cause further decline in certain salmon species (Goldberg, 1997; Edmonton Journal, 1997). If introduced GEOs foster increase supplies of desirable food fish resulting in lower prices, such instances of overfishing problems may be expected to intensify.

2.2.3 Negative ecological economic effects: an SMS approach

Introduced GEOs may also degrade ecosystem services that have high ecological economic value (e.g., biodiversity, global carbon cycling, detoxification, sequestration, etc.). Broad estimates of global ecosystem services have been attempted and marine ecosystem services are considered to provide the most economic benefits (Costanza et al., 1997b). However, lack of formal markets for ecosystem services makes estimating economic benefits, or economic costs of damage to ecosystem services from escaped GEOs, quite arduous. Although certain techniques exist for valuing non-market ecosystem services (see discussion of non-market benefit estimation

Economic Analysis of Introduced Genetically Engineered 197
Organisms

below), it may prove more efficient in the long-term to establish a Safe Minimum Standard (SMS) for natural services that may be at risk from introduced GEOs (Ciriacy-Wantrup, 1968; Bishop, 1978, 1979; Gowdy, 1997).

An SMS approach recognizes that future economic losses incurred from degrading ecosystem services resulting from introduced GEOs cannot be predicted with precision. When utilizing neoclassical and environmental economic approaches, there exists an "almost" adequate conceptual basis for economic valuation of ecosystem services such as biodiversity (Randall, 1988). However, Randall (1988) points out that due to the long-term nature of biodiversity as an asset, estimating benefits in present value terms at standard rates could lead to the inevitable collapse of living systems several hundred years from now, as relatively trivial economic gains in the immediate future outweigh discounted future benefits. Thus, a safe minimum standard could be established which maintains the overall structure and function of a specified marine ecosystem, supporting a similar abundance, diversity, and composition of the marine life and abiotic properties that existed prior to the introduction of the GEO.

Incorporating a safe minimum standard into the CBA implies that the maintenance of the current structure and function of marine systems is an economic good. The SMS decision rule is to maintain the current function and structure of natural marine systems (e.g., ecosystem services, biodiversity, etc.) unless the opportunity costs of doing so are intolerably high (Bishop, 1978). In short, the burden of proof for whether to allow alteration of natural systems from introduced GEOs is assigned to the case against maintaining the SMS. Implementation of SMS as a policy criterion may serve to reduce the risk of future economic ecosystem service losses that otherwise be result because of the uncertainty of outcomes inherent in marine GEO introductions.

2.3 Estimating Benefits and Costs of Marine GEOs: Empirical Methods

Upon review of the potential market and external effects from introduced GEOs, we offer some examples of methodologies that may be utilized to empirically estimate market and non-market benefits and costs from such introductions. First, we calculate one component of benefits derived from purchase of salmon in the market (consumer surplus). Next, we consider non-market benefit estimation and review commonly utilized empirical methods for valuing environmental goods and services that lack formal markets.

2.3.1 Estimating market benefits

Utilizing neoclassical empirical methods one can derive benefits from introduced GEOs by calculating consumer and producer surplus with given market prices and quantities. Here, we focus on benefits to consumers and derive the consumer surplus for one type of fish in the United States: salmon. (It is important to keep in mind, of course, that the analysis provided in this section is a simplified model offered for the purpose of illustration only. Lichtenberg et al. (1988) show that statistical linear demand curves, such as the one used here, do not necessarily provide accurate estimates of social surplus.) Table 1 contains mean production in metric tons and prices in thousands of dollars per metric ton for all salmon, aquacultured and wild, for the U.S., Canada, and Japan for the years 1984 to 1994. Using this economic data compiled by the FAO for commercial salmon (including aquacultured and natural fish), it is possible to estimate a simple demand curve for salmon in the U.S. By estimating the demand curve, one can obtain an empirical measure of consumer surplus.

Table 1. Quantities and values of Salmon, Selected Countries (Source: FAO, FSTAT, and AQUACULT databases, 1984-1994.

Country	Mean Annual Production (Metric Tons)	Average Price ($000s US per Metric Ton)	Mean Annual Aquaculture (Metric Tons)	Average Price ($000s US per Metric Ton)	Mean Wild Production (Metric Tons)	Average Price ($000s US per Metric Ton)
Canada	98,776	6.647	16,079	6.319	559,994	6.810
Japan	224,164	10.400	16,993	6.497	207,171	10.708
US	336,140	8.455	12,158	1.961	323,983	8.682

We statistically estimate two demand functions for Pacific and Atlantic salmon in the U.S. market: (1) demand for all salmon (i.e., wild plus aquacultured), and (2) demand for wild species only. The result of the first exercise produces a demand function of the form:

Quantity of Salmon Demanded = $625.92-28.79*Price of Salmon.

This equation can be presented graphically as shown in Figure 8, and CS can be calculated at any convenient price. In this case, we have chosen the

1994 price ($7.85 which is in thousands of dollars per metric ton, or about $3.57 per pound) to calculate annual benefits (CS per year for 1994). Note that by solving for the 1994 price, we can derive the quantity demanded that corresponds with the price. The consumer surplus calculated from this demand function is $2.78 billion, which is the benefit to consumers over and above the benefit obtained from consumption alone.

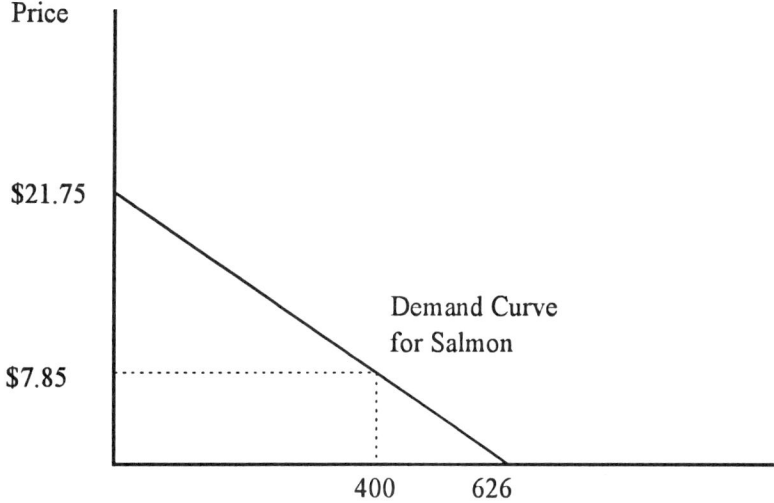

Figure 8. Estimated Consumer Surplus for Salmon in the U.S., 1994

A similar exercise was conducted for wild salmon using 1994 wild salmon prices ($8.27 in thousands of dollars per metric ton, or about $3.76 per pound) with an estimated demand function of:

Quantity of Wild Salmon Demanded = $646.13-31.98*Price of Salmon.

From this estimate, we obtain a CS measure of $2.28 billion implying CS for aquacultured salmon of $500 million. While aquacultured salmon in the U.S. are only 3.6% of production and less than 1% of the value of all salmon, the contribution by farmed salmon to consumer surplus is nearly 18%. This difference can be explained partially by the fact that increased production leads to decreased prices, so that more consumers are gaining surplus due to lower prices.

2.3.2 Estimating non-market benefits

In a CBA of introduced marine GEOs, utilization of implicit and direct pricing methods will be essential to internalize external effects and estimate the full social costs and benefits from an introduction. Economists often estimate values for non-market goods and services by utilizing implicit pricing methods. Utilization of implicit pricing methods to estimate values of environmental goods and services is extensive and increasingly applied in CBA (Hite et al., 1997; Hitzhusen, 1997; Randall, 1988). Certain implicit pricing methods are applicable when ecosystem goods and services can be purchased as complements or are characteristic of some ordinary marketed good (Randall, 1988). For example, environmental goods/services can be valued by assessing the travel costs that are incurred to enjoy the good (Clawson and Knetsch, 1966). In the case of introduced marine GEOs, one may estimate the travel costs of saltwater fisherman in an attempt to estimate benefits derived from the opportunity to enjoy recreational fishing in saline waters, undisturbed by introduced GEOs. Although travel costs would be case specific, expenditures can be substantial as indicated by the FWS study estimating total annual expenditures for U.S. saltwater anglers at $9.8 million (USFWS, 1991).

Another non-market valuation approach, hedonic pricing, involves decomposing housing prices to reveal a consumer's preference for associated environmental amenities and disamenities (e.g., scenic views, lakes, shorelines, etc.; Hite 1995; Hitzhusen et al., 1995). Any effects that lead to alteration of housing prices along coastlines may indicate the economic value of either enhancement or degradation of local marine systems from introduced GEOs. For example, algal blooms (e.g., red, green, or brown tides; see Chapter 4) may discolor water, lead to massive fish kills and may cause lost fishing opportunities as marine organisms become toxic for human consumption (Halverson and Martin, 1980; Norse, 1993; Villier, 1979). If introduced GEOs cause ecological effects which foster toxic phytoplankton blooms, one may observe local housing prices decline. Conversely, introduced GEOs that lead to the removal of such algal blooms through enhanced bioremediation should also be reflected in increased local housing prices.

Finally, contingent valuation methods involve direct valuation by surveying consumers and asking their willingness to pay (WTP) or willingness to accept (WTA) compensation for environmental goods and services in hypothetical situations (Randall et al., 1974; Randall, 1991). For instance, Green (1989) found that when redfish anglers were surveyed, they would be willing to pay between $5 and $14 per outing if they could increase their catch by 10%. Another example of this technique can be found in Smith and DesVouges (1985) in which estimates of the economic benefits of an improvement in water quality from boatable to fishable conditions ranged between $0.98 and $2.30 per outing (reported in 1981 dollars). Using such estimates, one can extrapolate across populations by multiplying average WTP or WTA of an individual by potential consumers. Thus, one can obtain an estimate of total benefits for environmental goods and services such as those derived from maintaining current levels of natural stocks of fish or a certain quality of water. In terms of introduced marine GEOs, one may assess the WTA compensation for degradation of current conditions (e.g., depletion of fish populations, decreased quality of water, etc.) or the WTP for maintenance of natural conditions (e.g., current status of fisheries) "secured" from escaped GEOs.

2.4 Regulatory and Containment Costs of Marine GEOs

The introduction of marine GEOs into production activities such as aquaculture will involve substantial investments, both private and public. Private costs that may be incurred include the cost of measures taken for containment and expenditures to conform to regulations such as sterilization of GEOs to avoid establishment in the wild. Costs of containment are variable and correlated to the effectiveness against escape. For example, prices per cubic meter for fish cages ranges from $10 to $100, with cages priced at $100/m3 being the most secure against escape (Forster, 1996). However, large expenditures for high levels of containment can make aquacultural activities unprofitable. Forster (1996) indicates that for offshore fish farming, cage prices above $50/m3 will probably not be profitable.

Due to perceived ecological and economic risks of transgenic organisms, containment measures may be more stringent and more costly. For example, field-testing of transgenic carp in outdoor ponds at Auburn University included fences, dikes, screens, 24-hour surveillance, and a system for applying poison in case of flooding (Fischetti, 1991). Further, the United States Department of Agriculture (USDA) concluded that installation of a so-called French drain in the outdoor transgenic fish ponds was warranted because this drainage system retains the smallest possible size of fish reared in the pond (ABRAC, 1995; USDA, 1990). In turn, increasing protection against escape requires increased production costs.

Containment may also include sterilization of GEOs to eliminate any possibility of establishment or spread of the transgene through hybridization upon escape. Sterilization techniques (e.g., triploid induction, eyestalk ablation, removal of gonadal tissue) will also result in higher production costs (ABRAC, 1995). Further, the efficacy of induced sterility varies greatly. For example, oyster cells have been reported to revert from a triploid state (i.e., infertile state) to the normal diploid state (Blankenship, 1994; also see Chapter 3).

In addition to the obvious costs to producers of meeting regulations on GEOs, it will fall on the government, and ultimately the taxpayer, to bear the costs of regulating introduced transgenic organisms in the marine environment. For instance, there will be costs associated with the implementation of safety protocols, including the costs associated with environmental risk assessment, costs of monitoring of aquaculture facilities to ensure that regulations are met, and legal costs for creating regulations and for court actions against noncompliance. In the U.S., total environmental regulatory costs are quite extensive; U.S. industry and government are currently spending about $115 billion a year to meet environmental standards set by legislation (estimates from the USGAO; Zilinskas et al., 1995). Legislation regulating introduced marine GEOs would increase regulatory costs, although the purpose would be to gain net social benefits by not degrading the valuable current structure and function of marine systems.

3. CONTAINMENT VS. POTENTIAL DAMAGES: AN ECONOMIC TRADE-OFF

In order to maximize net social benefits from introduced GEOs, one must determine the efficient level of expenditure that should be spent on containment measures in consideration of possible economic losses. An environmental economic model can be utilized as a framework for addressing this issue. Specifically, one can view the costs and benefits of introducing a GEO in terms of (1) marginal cost of avoiding or reducing a

Economic Analysis of Introduced Genetically Engineered Organisms

negative environmental outcome and (2) marginal damages to the environment caused by the outcome. In the case of marine GEOs, the negative environmental outcome, or damage, can encompass the economic risks outlined earlier in this chapter. To simplify, we assume here that the negative outcome caused by escape of a GEO is loss of biodiversity. The costs of avoiding biodiversity loss, (i.e., abatement costs), are incurred by the biotech industry (maintaining containment measures to avoid escape) and by government (monitoring the containment systems put in place by industry).

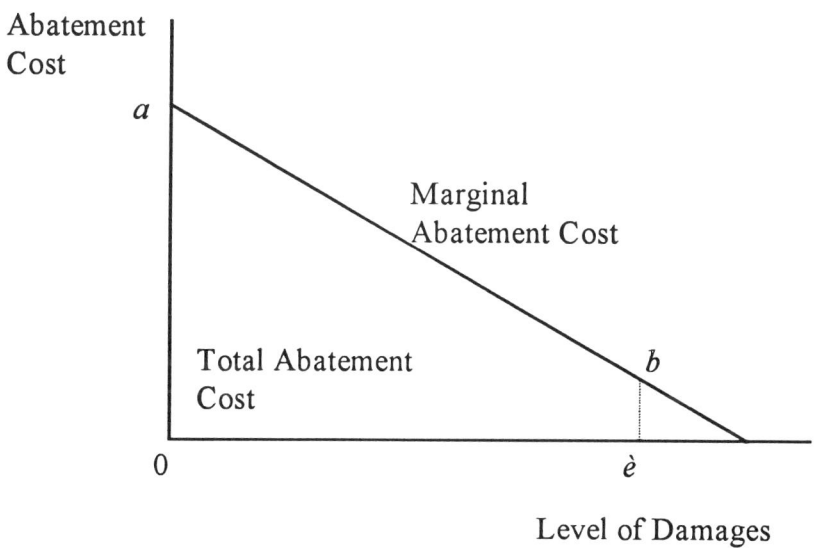

Figure 9. Abatement Costs of GEO Introductions

In Figures 9 and 10, we use the economic concepts of marginal abatement costs and marginal damages to graphically describe the costs and benefits of the introduction of GEOs. Specifically, Figure 9 illustrates the abatement costs associated with containment of a GEO. The vertical axis gives the per unit costs of abatement for the containment of GEOs. The horizontal axis shows the increasing damages associated with the escape of GEOs; note that at the point 0 on the graph, no damages would occur, and containment costs are extremely high. On the other hand, when damage levels are high (point è) abatement costs are low, but these costs increase when moving towards the origin. Total abatement costs from point è to point a are given by the total area under the marginal (i.e., per unit) abatement cost curve (MAC). Area abè0 is the cost of regulation for industry and government.

Figure 10 shows that damage (e.g., population decline, biodiversity loss, etc.) increases with greater numbers of GEOs released into the environment. The marginal damage cost curve (MD) gives the per unit cost of environmental economic damage associated with increased levels of released GEOs. The area aè0 gives the total cost of escaped GEOs at point è. Starting at point 0, there are no introductions of GEOs into the environment, but as escape increases, environmental costs increase as well. For instance, at low levels of escape, the impact may be localized, perhaps involving a small decline in a local fish population. However, with a large number of escapees damage may become substantial, such as a population crash or localized species extinction of an economically valuable foodfish (e.g., salmon) due to predation by an introduced GEO.

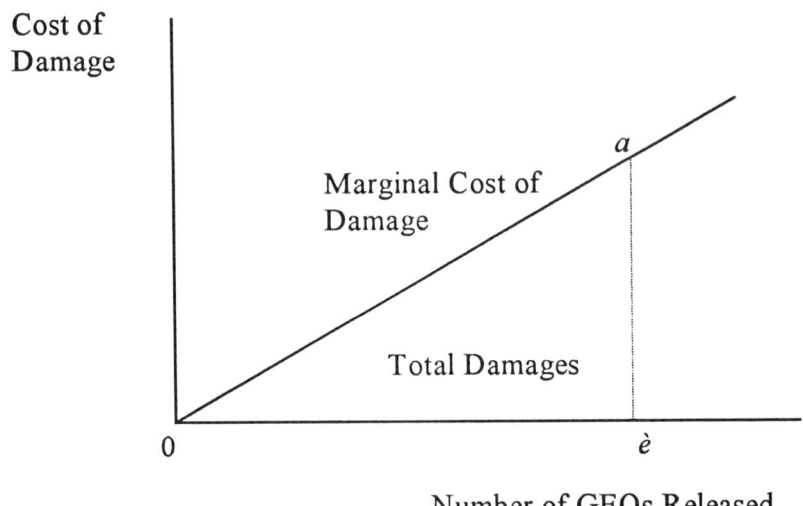

Figure 10. Number of GEOs Released

Upon estimation of containment costs and costs from environmental degradation, one can determine the efficient level of containment to be utilized for GEOs. In Figure 11, the efficient amount of containment occurs at the point at which marginal abatement costs equal marginal damages. In

Economic Analysis of Introduced Genetically Engineered Organisms

this model, release of more GEOs than the equilibrium quantity results in a condition in which marginal damages would exceed marginal containment costs and thus degradation of the environment occurs. Conversely, escape of few GEOs, less than the equilibrium quantity, indicates that there has been too much expenditure on containment measures because further probable damage from the GEOs is less costly than expenditures for such extensive containment. Obviously, if it is determined that any release of GEOs will lead to substantial costs, the marginal damage curve may occur above the marginal abatement cost curve (with no intersection) and thus indicate that the allowable number of released GEOs is zero. In summary, within a CBA framework, once estimates have been established for potential economic damages from introduction of GEOs and containment costs have been determined, analysis with MAC and MD curves serves as a useful approach for determining efficient levels of containment.

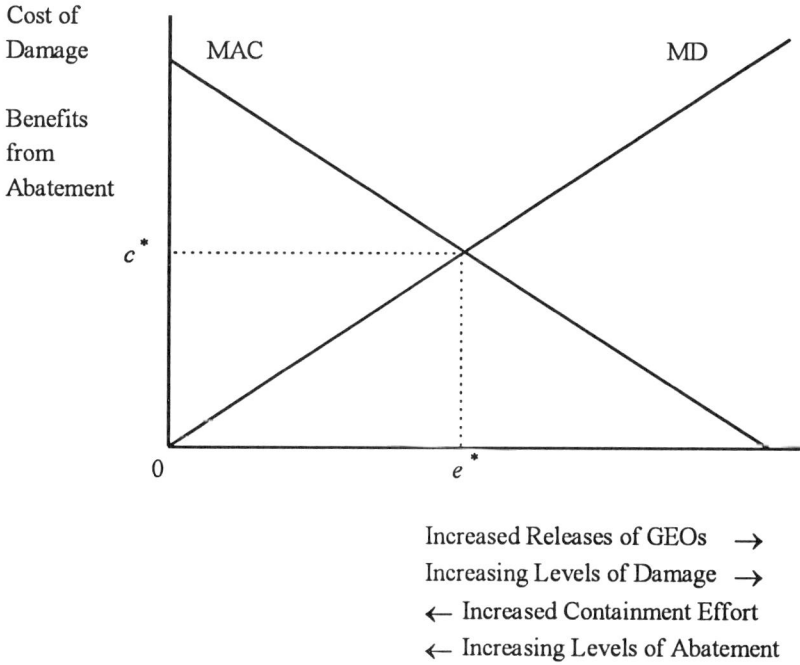

Figure 11. Efficient Level of Containment for Introduced Marine GEOs

4. CONCLUSION

In this chapter, we have provided frameworks, policy criteria, and empirical methodologies for analyzing economic effects of introduced marine GEOs. Further, we have identified potential intramarket and external costs and benefits of introduced marine GEOs. To summarize, introduced marine GEOs may offer substantial social benefits by meeting the growing global demand for seafood, acting indirectly to maintain valuable natural marine stocks, and fostering restoration attempts as enhanced transgenic bioremediators. However, potential economic costs include degradation of natural and recreational fisheries from escaped GEOs, indirect perverse incentives that lead to overfishing, and direct degradation of production activities from biofouling activities. Thus, for each introduction a case-specific cost-benefit analysis (CBA) will be required to determine whether the effect of introduced marine GEOs provides net societal benefits.

Here, we have promoted the utilization of an economic cost-benefit analysis to augment the risk assessment process, but not replace it. Interdisciplinary collaboration between ecologists and economists will be essential to obtain monetary estimates for environmental effects that can be valued with current empirical economic methods. Further, ecological services beyond the realm of current economic valuation techniques can be integral to the sustainable functioning of marine systems and thus must also be considered in the risk assessment process utilizing other metrics. Implementation of a safe minimum standard may be warranted to ensure that valuable ecosystem services, that are difficult to value with current economic techniques, shall be maintained at functioning levels.

Economic evaluation of introduced marine GEOs will require a case-by-case approach in order to determine relevant factors and the breadth of the cost-benefit analysis. Once benefits and costs of introduced marine GEOs have been estimated, the amount of efficient expenditure for containment measures can be determined by considering potential damages versus abatement costs. Logically, one would expect regulatory and compliance costs to arise from the establishment and enforcement of a risk assessment protocol for introduced marine GEOs. However, an efficient risk assessment protocol will maximize net social benefits by safeguarding natural marine systems if degradation proves to be more costly than containment measures. In the risk assessment process, integration of economic analyses with findings of potential environmental effects can serve to quantify certain risks from introduced marine GEOs in monetary terms and provide useful information for effective decision making.

REFERENCES

Agricultural Biotechnology Research Advisory Committee (ABRAC). 1995. Performance Standards for Safely Conducting Research with Genetically Modified Fish and Shellfish. Washington: United States Department of Agriculture (USDA).

Aquatic Species Nuisance Task Force (ASNTF). 1992. Proposed Aquatic Nuisance Species Program. Alexandria, VA: U.S. Fish and Wildlife Service.

Barbier, E.B. 1989. Economics, Natural-Resource Scarcity and Development. London: Earthscan Publications Ltd.

Barbier, E.B., J.C. Burgess and C. Folke. 1994. Paradise lost? The ecological economics of biodiversity. London: Earthscan.

Bishop, R.C. 1978. Endangered Species and Uncertainty: The Economics of a Safe Minimum Standard. American Journal of Agricultural Economics 60(1):10-18.

Bishop, R.C. 1979. Endangered Species, Irreversibility and Uncertainty: A Reply. American Journal of Agricultural Economics 61(2):376-379.

Blankenship, K. 1994. Experiment with Japanese oyster ends abruptly: oysters thought to be sterile found capable of reproducing. Bay Journal 4(5):1-4. Baltimore, MD: Alliance for Chesapeake Bay.

Botsford, L.W., J.C. Castilla and C.H. Peterson. 1997. The Management of Fisheries and Marine Ecosystems. Science 277:509-514.

Cairns, J. Jr and J.R. Pratt. 1986. Ecological consequence assessment: effects of bioengineered organisms. In: J. Fiskel and V.T. Covello (eds.). Biotechnology Risk Assessment: Issues and Methods for Environmental Introductions. New York: Pergamon Press.

Chen, T.T., J-K. Lu, M.J. Shamblott, C.M. Cheng, C-M. Lin, J.C. Burns, R. Reimschuessel, N. Chataknondi and R.A. Dunham. 1996. Transgenic fish: ideal models for basic research and biotechnological applications. In: J.D. Ferraris and S.R. Palumbi (eds.). Molecular Zoology. New York: John Wiley and Sons, Inc.

Christensen, J. 1997. Cultivating the world's demand for seafood. New York Times, March 1st: 27-29.

Ciracy-Wantrup, S.V. 1968. Resource Conservation: Economics and Policies. 3rd ed. Berkeley, CA: University of California Press.

Clawson, M. and J. Knetsch. 1966. Economics of Outdoor Recreation. Baltimore: Johns Hopkins University Press.

Clayton, M. 1995. A Fish Tale? Canada Tries to Save Stocks While Overfishing. Christian Science Monitor. March 20th.

Cohen, A.N. and J.T. Carlton. 1995. Non-Indigenous Aquatic Species in a United States Estuary: A Case Study of the Biological Invasions of the San Francisco Bay and Delta. Report to the U.S. Fish and Wildlife Service, Washington, DC, December, 1995.

Colwell, R. 1987. Genetically engineered organisms in the ocean environment -risks and benefits. In: M.A. Sleigh (ed.). Microbes in the Sea. New York: John Wiley and Sons.

Costanza, R., J. Cumberland, H. Daly, R. Goodland and R. Norgaard. 1997a. An Introduction to Ecological Economics. Boca Raton, FL: CRC Press.

Costanza et al., 1997b. The value of the world's ecosystem services and natural capital. Nature 387:253-260.

Costanza, R. (ed.). 1991. Ecological Economics: The Science and Management of Sustainability. New York: Columbia University Press.

Daily, G.C. (ed.). 1997. Nature's Services: Societal Dependence on Natural Ecosystems. Washington: Island Press.

Daly, H.E. 1991. The elements of environmental macroeconomics. In: R. Costanza (ed.). Ecological Economics: The Science and Management of Sustainability. Columbia University Press: New York.

Dasgupta, A.K. and D.W. Pearce. 1978. Cost-Benefit Analysis: Theory and Practice. London: Macmillian Press, Ltd.

Devlin, R.H., T.Y. Yesaki, C.A. Blagi, E.M. Donaldson, P. Swanson and C. Woon-Khlong. 1994. Extraordinary salmon growth. Nature 371:209-210.

Dixon, J.A., L.F. Scura, R.A. Carpenter and P.B. Sherman. 1994. Economic Analysis of Environmental Impacts. London: Earthscan Publications, Ltd.

Dixon, J.A., L.F. Scura and T. van't Hof. 1993. Meeting Ecological and Economic Goals: Marine Parks in the Caribbean. AmBio 23(2-3):117-125.

Dixon, J.A. and M.M. Hufschmidt. 1986. Economic Valuation Techniques for the Environment. Baltimore: Johns Hopkins University Press.

Edmonton Journal. 1997. Canadians get most fish. Edmonton Journal Extra National News. Internet News Archive. August 9.

Ehrlich, P.R. and A.H. Ehrlich. 1992. The value of biodiversity. Ambio 21:219-226.

Field, B.C. 1994. Environmental Economics: An Introduction. New York: McGraw Hill, Inc.

Fischetti, M. 1991. A Feast of Gene-Splicing Down on the Fish Farm. Science 253:512-513.

Food and Agriculture Organization of the United Nations (FAO). 1996. Fishstat Database. Rome: FAO.

Food and Agriculture Organization of the United Nations (FAO). 1995a. Review of the State of the World Fishery Resources: Marine Fisheries. FAO Circular No. 884. Rome: FAO.

Food and Agriculture Organization of the United Nations (FAO). 1995b. The State of the World Fisheries and Aquaculture. Rome: United Nations.

Food and Agriculture Organization of the United Nations (FAO). 1995c. Global Fish and Shellfish Production in 1993. FAO Fisheries Department, Fisheries Information, Data and Statistics Service. Rome: FAO.

Food and Agriculture Organization of the United Nations (FAO). 1993. Marine Fisheries and the Law of the Sea: A Decade of Change. FAO Fisheries Circular No. 853. Rome: FAO.

Forster, J. 1996. Cost and Market Realities in Open Water Aquaculture. Proceedings of the Open Ocean Aquaculture Conference, May 8-10, 1996.

Frederick, R.J. and M. Egan. 1994. Environmentally compatible applications of biotechnology. BioScience 44(8):529-535.

Garcia, S. and C. Newton. Current Situation, Trends, and Prospects in World Capture Fisheries. 1994. Paper presented at the Conference on Fisheries Management: Global Aspects. Seattle, Washington.

Garrett, L. 1994. The Coming Plague: Newly Emerging Diseases in a World Out of Balance. New York: Farrar, Straus and Giroux.

Goldberg, C. 1997. Fishing Talks and Fighting Words. New York Times June 5th p16.

Gowdy, J.M. 1997. The Value of Biodiversity: Markets, Society, and Ecosystems. Land Economics 73(1):25-41.

Green, T. 1989. The Economic Value and Policy Implications of the Recreational Red Drum Success Rates in the Gulf of Mexico. Report to the National Marine Fisheries Service, University of Southern Mississippi, July, 1989.

Gutrich, J.J. 1997. Biodiversity and Developing Countries. In: F.J. Hitzhusen (ed.). Problems and Policies in World Population, Food and Environment: An Integrated Approach. Columbus, OH: The Ohio State University.

Halverson, M. and D.F. Martin. 1980. Studies of cytoclysis of *Chattonella subsalsa*. Florida Science 43(1):35.

Hite, D. 1995. Estimation of Welfare Impacts of an Environmental Disamenity in the Residential Real Estate Market. PhD Dissertation. The Ohio State University, Department of Agricultural Economics. Columbus, OH.

Hite, D., L.D. Forster and J. Rausch. 1997. Optimal Use of Flue Gas Desulfurization By-Product. Selected Paper, Presented at the 1997 Annual Meeting of the American Agricultural Economics Association. July 27-30th. Toronto, Canada.

Hitzhusen, F.J. 1997. Economic Analysis of Sedimentation Impacts for Watershed Management. Paper presented at the International Symposium: Global Challenges in Ecosystem Management in a Watershed Context. July 25-26th. Toronto, Canada.

Hitzhusen, F.J. et al. 1995. Economics and Political Analysis of Dredging Ohio's State Park Lakes. In: A. Divar and E.T. Lochman (eds.). Water Quantity/Quality Management and Conflict Resolution: Institutions, Processes and Economic Analyses. Westport, CT: Praeger.

Hitzhusen, F.J. 1984. Cost-Benefit Analysis: Cornerstone or Achilles' Heel of Social Science? Presentation at the Social Science Colloquium, East-West Resource Systems Institute, University of Hawaii.

Infofish. 1995. Heavier fish through genetic engineering. INFOFISH International 2:72.

Jansson, A., M. Hammer, C. Folke and R. Costanza. 1994. Investing in Natural Capital. Washington: Island Press.

Joint Subcommittee on Aquaculture (JSA). 1992. Aquaculture in the United States: Status, Opportunities and Recommendations. A Report to the Federal Coordinating Council on Science, Engineering and Technology. Washington: United States Department of Agriculture.

Kaufman, L. and P. Dayton. 1997. Impacts of Marine Resource Extraction on Ecosystem Services and Sustainablity. In: G.C. Daily (ed.). Nature's Services: Societal Dependence on Natural Ecosystems. Washington: Island Press.

Kane, H. 1993. Growing Fish in Fields. WorldWatch 6:5:20-27.

Krishnan, R., J.M. Harris, and N.R. Goodwin (eds.). A Survey of Ecological Economics. Washington: Island Press.

Lichtenberg, E., D. Parker and D. Zilberman. 1988. Marginal analysis of welfare costs of environmental policies. American Journal of Agricultural Economics

Lipton, D.W., K. Wellman, I.C. Sheifer and R.F. Weiher. 1995. Economic Valuation of Natural Resources: A Handbook for Coastal Resource Policymakers. NOAA Coastal Ocean Program Decision Analysis Series No. 5. Silver Spring, MD: NOAA Coastal Ocean Office.

Lundin, C.G. and R.A. Zilinskas. 1995. Marine Biotechnology in the Asian Pacific Region. Washington: The World Bank.

MacKensie, D. 1996. Can we make supersalmon safe? New Scientist January 27:14-15.

Milon, J.W. and J.F. Shogren (eds.). 1995. Integrating Economic and Ecological Indicators: Practical Methods for Environmental Analysis. London: Praeger.

National Science and Technology Council (NSTC; Office of Science and Technology Policy). 1995. Biotechnology for the 21st Century: New Horizons. National Science and Technology Council. Washington: U.S. Government Printing Office.

Norgaard, R. 1988. The Rise of the Global Exchange Economy and the Loss of Biological Diversity. In: E.O. Wilson and F.M. Peter (eds.). Biodiversity. Washington: National Academy Press.

Norse, E.A. (ed.). 1993. Global Marine Biological Diversity: A Strategy for Building Conservation into Decision Making. Washington: Island Press.

Parker, N. 1992. Economic pressures driving genetic changes in fish. In: A. Rosenfield and R. Mann (eds.). Dispersal of Living Organisms Into Aquatic Organisms. College Park, MD: Maryland Sea Grant College.

Pearce, D. 1985. Sustainable Futures: Economics and the Environment. Inaugural Lecture: University College London, December 5th, 1985

Peterson, C.H. and J. Lubchenco. 1997. Marine Ecosystem Services. In: G.C. Daily (ed.). Nature's Services: Societal Dependence on Natural Ecosystems. Washington: Island Press.

Powers, D. 1995. New Frontiers in Marine Biotechnology: Opportunities for the 21st Century. In: C.G. Lundin and R.A. Zilinskas. Marine Biotechnology in the Asian Pacific Region. Stockholm: The World Bank.

Randall, A. 1991. Total and Nonuse Values. In: J.B. Braden and C. D. Kolstad (eds.). Measuring the Demand for Environmental Quality. New York: Elsevier Science Publishers.

Randall, A. 1988. What Mainstream Economists Have to Say About the Value of Biodiversity. In: E.O. Wilson and F.M. Peter (eds.). Biodiversity. Washington: National Academy Press.

Randall, A. 1987. Resource Economics: An Economic Approach to Natural Resource and Environmental Policy. 2nd Edition. New York: John Wiley and Sons.

Randall, A., B. Ives and C. Eastman. 1974. Bidding Games for Valuation of Aesthetic Environmental Improvements. Journal of Environmental Economics and Management 1(2).

Rapport, D.J. and J.E. Turner. 1977. Economic models and ecology. Science 195:367-373.

Rothschild, M. 1990. Bionomics: Economy As Ecosystem. New York: John MacRae.

Scura, L.F. and T. van't Hof. 1993. Economic Feasibility and Ecological Sustainability of the Bonaire Marine Park. Environment Department Divisional Working Paper 1993-44. Washington: The World Bank.

Shears, M.A., G.L. Fletcher, C.L. Hew, S. Gauthier and P.L. Davies. 1991. Transfer, expression and stable inheritance of antifreeze proteins in Atlantic salmon. Molecular Marine Biology and Biotechnology 1(1): 58-63.

Smith, V., Kerry and W. DesVousge. 1985. The Generalized Travel Cost Model and Water Quality Benefits: A Reconsideration. Southern Economic Journal 52:371-381.

Tietenberg, T. 1996. Environmental and Natural Resource Economics. 4th Edition. New York: HarperCollins

Tiedje, J.M., R.K. Colwell, Y.L. Grossman, R.E. Hodson, R.E. Lenski, R.N. Mack and P.J. Regal. 1989. The planned introduction of genetically engineered organisms: Ecological considerations and recommendations. Ecology 70:298-315.

Travis, J. 1993. Invader threatens Black, Azov seas. Science 262:1366-1367.

United States Department of Agriculture (USDA). 1990. Environmental Assessment of Research on Transgenic Carp in Confined Outdoor Ponds to be Conducted at the Alabama Agricultural Experiment Station (AAES), Auburn University, Auburn, Alabama. Washington: USDA.

United States Fish and Wildlife Service (FWS). 1988. 1985 National Survey of Fishing, Hunting and Wildlife Associated Recreation. Washington: U.S. Department of Interior.

United States Fish and Wildlife Service (FWS). 1993. 1991 National Survey of Fishing,Hunting and Wildlife Associated Recreation. Washington: U.S. Department of Interior.

Villier, G. 1979. Recovery of a population of white mussels *Donax serra* at Elands Bay, South Africa, following a mass mortality. Fisheries Bulletin (South Africa. Sea Fisheries Branch) 12:69-74.

Weber, P. 1995. Protecting oceanic fisheries and jobs. In: State of the World 1995. New York: W.W. Norton and Co.

World Resources Institute (WRI). 1996. World Resources Guide to the Global Environment. New York: Oxford University Press.

Zilinskas, R.A., R.R. Colwell, D.W. Lipton and R.T. Hill. 1995. The Global Challenge of Marine Biotechnology. College Park, MD: Maryland Sea Grant College.

Chapter 7

Risks and Benefits of Marine Biotechnology: Conclusions and Recommendations

PETER J. BALINT, RITA R. COLWELL, JOHN J. GUTRICH, DIANE HITE, MORRIS LEVIN, SUSAN STENQUIST, HOWARD H. WHITEMAN, AND RAYMOND A. ZILINSKAS

1. INTRODUCTION

Marine biotechnology already provides substantial social benefits, including improved productivity in aquaculture, new pharmaceuticals, and enhanced manufacturing with enzymes derived from marine animals, plants, and microorganisms. A broad-based industry is developing that will bring economic gains to entrepreneurs, their employees, the communities in which they establish their facilities, and indeed to the entire nation.

Expanded research and development in genetic recombination of marine organisms is leading to beneficial discoveries with wide application. A "blue" revolution in aquaculture may be on the horizon, following the model of the "green" revolution in agriculture, in which new techniques create faster-growing, more nutrient-efficient fish and shellfish to provide a source of protein for the world's expanding human population while reducing harvest pressures on natural fisheries.

Marine biotechnology is also poised to make an important contribution in the area of bioremediation. To help reverse decades of environmental degradation in bays, estuaries, and other fragile coastal ecosystems, the techniques of modern molecular biology are being applied to enhance the capabilities of naturally occurring microorganisms that are effective metabolizers of sulfur, nitrogen, and hydrocarbons.

As with any new technology, however, all aspects of proposed applications need to be considered, including potential risks. Some risks are clearly understood and easy to manage; others may be difficult to foresee or entirely unknown. In particular, it is hard to predict the ecological consequences of introducing GEOs into marine environments. Containment of introduced organisms may be impossible, and escaped transgenic fish or bacteria may establish viable populations in the wild and alter natural ecosystems.

Other considerations are social and economic. Extended delays or overly burdensome compliance procedures will result if regulatory agencies are unprepared to provide effective and efficient evaluations of permit applications because of a lack of basic scientific knowledge or a lack of appropriate risk assessment protocols. Oversight procedures must be designed to minimize adverse environmental effects while maximizing social and economic benefits. Moreover, the general public must have the opportunity to consider associated cultural and ethical issues before new products come to market. Unlike other high technology applications that alter society in fundamental ways—advances in computer applications, for example—biotechnology affects life processes. If adequate protections are not assured, unfounded fears of biological change may undercut commercial viability of beneficial products.

This volume explores the risks and benefits of marine biotechnology and considers appropriate policy responses. Contributors assess the level of scientific knowledge required for ecological risk assessment of genetically engineered marine organisms, propose a model for ecological analysis that takes scientific uncertainty into account, make recommendations for legislative and regulatory reform, and evaluate the economic implications of marine biotechnology. Here, the authors present a brief summary of conclusions and recommendations drawn in preceding chapters.

2. THE SCIENTIFIC KNOWLEDGE BASE

Marine organisms interact with their environment in ways that are significantly different and less well understood than terrestrial species. Even though continuing research is narrowing gaps in the knowledge base, it will be more difficult to predict the nature and extent of ecological effects in the marine environment than has been the case in terrestrial applications. Introduced marine GEOs will be more difficult to contain and retrieve, and organisms that do escape containment will have wider opportunities for reproduction and dispersal. The majority of plants and animals used in agriculture that have been modified through genetic engineering are domesticated species dependent upon human mediation for survival, and the risk that they will establish viable fugitive populations and disturb natural ecosystems is limited. In contrast, genetically engineered fish, shellfish, and marine microorganisms will be more difficult to isolate from the environment once introduced and will retain their natural survival skills if they escape containment. Furthermore, the effects of an escape may not be localized. Many marine organisms utilize long distance dispersal mechanisms to establish populations in widely separated regions of the oceans.

These factors increase the uncertainty inherent in ecological risk assessment in the marine environment. Although it is unlikely that marine GEOs pose a direct threat to human health, the level of risk posed to natural ecosystems is unclear. Experience with exotic species indicates that in general there is minimal impact when organisms are introduced into novel habitats. Most organisms are adapted to their natural environment and demonstrate reduced fitness in other locations. In cases where adverse outcomes do occur, however, negative ecological and economic effects can be dramatic and widespread. In response, researchers are developing ways to modify GEOs to minimize the likelihood of adverse effects.

The need for further research in marine ecology and ecosystem function is the major obstacle to full exploitation of the benefits of marine biotechnology. At a minimum, research targeting the following issues should be accelerated: microbial ecology of aquaculture systems, interactions of transgenics with local communities and food webs, and effects of transgenic manipulation on life history traits.

Incomplete information also delays regulatory approval. Currently, risk assessors must consider proposals to develop and test marine GEOs cautiously, on a case-by-case basis. Government and industry should support increased basic research to lay the groundwork for an approval process that would be efficient and responsive, while maintaining high standards of environmental protection.

3. A MODEL FOR ECOLOGICAL ANALYSIS

In the conventional view, introduced organisms are said to have "escaped" if they are found outside the containment area or the area of uncontained introduction. To conduct an analysis of ecological effects that may be associated with marine GEOs, however, a broader definition is required. For example, if introduced GEOs utilize novel resources unavailable to the parent organisms (e.g., new prey or refugia) this represents an escape from existing ecological constraints even though the altered organisms remain within the local ecosystem. Escape must be defined to include any change in the fundamental niche of the parental strain.

As a first step in identifying ecological risk, regulators must assess the probability of escape using this broader definition. Next, they must estimate the probability that escaped GEOs will successfully establish viable, fugitive populations. Although there may be short-term effects caused by nonreproductive individuals, a higher level of scrutiny is required for organisms that are reproductive, or that have the potential to revert to reproductive status from induced sterility. Third, risk assessors must evaluate the rate at which fugitive populations may establish themselves, and, once established, spread across a habitat or among habitats. Fourth, they must determine the overall likelihood of ecological impact, given their estimates of the parameters listed above. Finally, decision makers must weigh potential risks against expected benefits.

Although ecological risk varies on a case-by-case basis, we support the following general assessment in this book: the likelihood of escape for introduced marine GEOs must be considered high; the risk that escaped organisms will establish viable populations is moderate; the rate of establishment of fugitive populations within enhanced niches is likely to be high; and the rate of spread of fugitive populations across habitats is likely to be moderate, although rapid short-term expansion is possible. In summary, we consider the over-all potential for ecological effects in marine ecosystems colonized by populations of escaped GEO to be moderate. As mentioned above, however, the level of risk can be reduced significantly if GEOs are modified to restrict their opportunities for long-term survival, reproduction, and dispersal.

4. LEGISLATIVE AND REGULATORY REFORM

Legislation currently in place does not specifically address marine GEOs, and there is no clear statutory mandate for oversight of these products as they are developed for commercial applications. Presently, genetically engineered microorganisms are regulated by the Environmental Protection Agency (EPA) under authority granted by the Toxic Substances Control Act (TSCA). Food animals intended for human consumption, including marine aquacultural harvests, are regulated by the Food and Drug Administration (FDA) under authority granted by the Federal Food, Drug and Cosmetic Act (FFDCA). By default, ecological risk assessment is the responsibility of the agency with primary human health jurisdiction. As a result, EPA evaluates ecological risk for transgenic marine microorganisms, and FDA assesses ecological risk for genetically engineered fish and shellfish.

Because standardized protocols have not been established, FDA personnel oversee development of transgenic fish using guidelines promulgated in 1995 by the now-defunct Agricultural Biotechnology Research Advisory Committee (ABRAC) of the U.S. Department of Agriculture (USDA). The ABRAC documents provide a pathway of over 20 detailed flowcharts and worksheets designed to ensure safety and security during research and development of genetically engineered aquaculture products. Potential ecological risks are also considered. In our view, the primary drawback of the ABRAC risk management methodology lies in its unnecessary specificity. There is little room for case-by-case distinctions and the use of best professional judgment. In contrast, the EPA list of points to consider and the USDA biotechnology products application documents lay out a more general framework of basic questions to be asked that allow regulators to draw from experience to determine the level of scrutiny applicable in each case.

To protect the environment—in addition to safeguarding human health and ensuring food safety—risk assessment must include input from experts in marine biotechnology, oceanography, marine ecology, and the ecological dynamics of marine fisheries. In reviewing applications to introduce GEOs into the marine environment, regulators at EPA and FDA should collaborate with personnel from outside agencies, including the National Oceanographic and Atmospheric Administration (NOAA), the Fish and Wildlife Service (FWS), the National Marine Fisheries Service

(NMFS), and the National Sea Grant Program. If possible, an inter-agency oversight group should be established.

According to the Federal Coordinated Framework for the Regulation of Biotechnology, established in 1986, biotechnology products should be evaluated based on their expressed characteristics, not the tools used in their production. At present, however, regulators implicitly, and often explicitly, consider the production process, thereby applying different standards to biotechnology products than to conventional products. Furthermore, regulations and procedures vary from agency to agency. These regulatory inconsistencies, and other weaknesses in the risk assessment structure, may slow development of beneficial products and retard economic progress in the marine biotechnology industry. Although statutory clarification would be useful, Congressional action is not required. Established protocols, including the EPA Points to Consider, can be interpreted broadly enough to ensure appropriate appraisal of the risks associated with marine GEOs, provided the appropriate expertise is brought to bear. If the recommendations outlined here are implemented, including accelerated basic research and standardization of oversight procedures, policy makers and regulators will be prepared to act expeditiously; if not, unnecessary delays will occur as government agencies strive to catch up to private sector advances.

To complicate matters, non-federal government agencies also have oversight obligations. The national government has authority over U. S. territorial waters and regulates activities conducted in the Exclusive Economic Zone beyond the three-mile limit. Inside that demarcation, however, regional, state, and local agencies may exercise oversight responsibility. Federal and state agencies often do not have standardized procedures and do not cooperate efficiently. In addition, many coastal states with an interest in regulating commercial enterprises related to marine biotechnology are not prepared to undertake such supervision. The necessary statutes have not been enacted and the appropriate expertise is not available within their agencies.

Policy makers must also consider international implications. For economic reasons related to global competition, and for ecological reasons associated with the dispersal characteristics and life histories of marine organisms, adoption of multilateral agreements is an important component of any effort to manage risk comprehensively. Developed nations need to standardize responses to products of marine biotechnology, and developing nations need to maintain environmental protection even as they work to attract private investment.

5. ECONOMIC IMPLICATIONS

Costs to the marine biotechnology industry associated with regulatory compliance cannot yet be estimated with accuracy. Risk assessment protocols have not been standardized, nor is it clear which offices of which agencies will exercise federal, state, and local oversight authority. Industry may have to comply separately with procedures administered by various agencies with differing agendas, levels of expertise, and institutional cultures. As a result, estimates of compliance costs range across a wide spectrum.

If regulations are appropriately applied, however, costs to industry may not be exorbitant. The potential for direct impact on human health and food safety from biotechnology-based marine aquaculture or bioremediation is small. As a result, marine biotechnology companies should not have to meet costs presently borne, for example, by firms attempting to introduce new pesticides or food additives. Satisfying regulatory requirements for these products can require expenditures in the millions of dollars; extensive toxicological studies using animal models are often necessary. In contrast, the cost of ascertaining that transgenic fish and shellfish are safe for human consumption, and receiving agency approval for production and marketing based on these criteria, may have associated costs that are relatively modest.

Industry may find that significantly larger outlays are required to bolster the inadequate science base presently available to regulators attempting to estimate ecological risk. Private companies, particularly those first in line with products requiring approval, may find it necessary to fund their own research in marine ecology and the effects of genetic manipulation on morphology, life history, and fitness to provide data to support their applications. This may pose a substantial burden. The costs of supporting such studies can be high both in terms of time and money. A single project, targeting a limited question, may take a year or more to complete and may require an investment of several hundred thousand dollars. Many such studies are needed. When added to in-house research and development costs, and the outlays for public relations and advertising that marketing such products to a wary public will demand, it is clear that private development of marine biotechnology products is a capital intensive undertaking. Potential profit margins may be such that additional costs associated with conducting large-scale basic research would be an insupportable burden. In addition, industry is concerned with the appearance of conflict of interest in studies funded by companies with an economic stake in the results.

This logic underscores the importance of increasing government-sponsored research. The investment in public funds will pay dividends as beneficial products come to market and the marine biotechnology industry contributes to international economic competitiveness. If an appropriate regulatory framework is established, supported by adequate scientific data, compliance costs to industry will be manageable, the environment will be protected, and the nation will reap substantial social and economic benefits.

6. THE FUTURE OF MARINE BIOTECHNOLOGY

The first successful gene transfer in fish occurred in 1986. Scientists isolated the segment of DNA that codes for expression of growth hormone production in rainbow trout and transferred it into the common carp. The result was a fish that grew up to 40% faster and reached a larger size than its unaltered siblings. That same year the first agricultural biotechnology product was field-tested. Researchers demonstrated that the so-called "ice-minus" bacteria could perform the task for which they were designed—to retard the formation of frost on strawberry plants.

Since that time, firms have brought many pharmaceutical and agricultural biotechnology products to market. *E. coli* bacteria now produce human insulin, and cabbage looper (*Trichoplusia ni*) larvae manufacture interleukin-2. In agriculture, farmers plant herbicide-resistant cotton and corn, and raise livestock, including cattle and pigs, with enhanced milk production and meat quality.

Until recently, marine biotechnology has progressed more slowly. Now the pace is accelerating as industry prepares a new generation of applications for use in marine aquaculture and bioremediation. Policy makers can encourage growth in this emerging industry, and speed the development of beneficial, environmentally safe products, by reforming oversight procedures and investing in basic research to realize the dual goal of economic growth and environmental protection.

LIST OF ACRONYMS

ABRAC	Agricultural Biotechnology Research Advisory Council
ACRE	Advisory Committee on Release to the Environment (UK)
AFS	American Fisheries Society
ANS	Aquatic Nuisance Species
ANSTF	Aquatic Nuisance Species Task Force
API	American Petroleum Institute
BSCC	Biotechnology Science Coordinating Committee
Bt	*Bacillus thuringiensis*
CBA	Cost-Benefit Analysis
CBP	Chesapeake Bay Program
CFR	Code of Federal Regulations
CGLF	Convention on Great Lakes Fisheries
CLF	Conservation Law Foundation
CMMB	Committee on Molecular Marine Biology (NRC)
COE	U.S. Army Corps of Engineers
CNI	Community Nutrition Institute
CPIB	Center for Public Issues in Biotechnology
CS	Consumer Surplus
CVM	Center for Veterinary Medicine
CWA	Clean Water Act
CZMA	Coastal Zone Management Act
DOC	U.S. Department of Commerce
DOI	U.S. Department of Interior
DNA	Deoxyribonucleic Acid
EA	Environmental Assessment
EEZ	Exclusive Economic Zone
EIS	Environmental Impact Statement
ENSO	El Niño Southern Oscillation Current
EPA	U.S. Environmental Protection Agency
FAO	Food and Agriculture Organization of the United Nations
FDA	U.S. Food and Drug Administration
FFDCA	Federal Food, Drug, and Cosmetic Act
FIFRA	Federal Insecticide, Fungicide, and Rodenticide Act
FR	Federal Register
FSIS	Food Safety and Inspection Service
FWCA	Fish and Wildlife Coordination Act
FWS	U.S. Fish and Wildlife Service

GEM	Genetically Engineered Microorganism
GEMB	Genetically Engineered Marine Bacteria
GEMM	Genetically Engineered Marine Microorganism
GEO	Genetically Engineered Organism
GLWQA	Great Lakes Water Quality Agreement
ICES	International Committee for the Exploration of the Seas
IJC	International Joint Commission
ITFB	Interagency Task Force on Biotechnology
JSA	Joint Subcommittee on Agriculture
MAC	Marginal Abatement Cost
MCAN	Microbial Commercial Activities Notification
MD	Marginal Damage
NAL	National Agricultural Library
NAS	National Academy of Sciences
NBIAP	National Biological Impact Assessment Program
NDBAP	New Developments in Biotechnology Advisory Panel
NEPA	National Environmental Policy Act
NIH	National Institutes of Health
NMFS	U.S. National Marine Fisheries Service
NOAA	National Oceanographic and Atmospheric Administration
NPDES	National Pollution Discharge Elimination Systems
NRC	National Research Council
OECD	Organization for Economic Cooperation and Development
OPP	Office of Pesticide Programs
OPPTS	Office of Pesticides, Pollution and Toxic Substances
OSTP	Office of Science and Technology Policy
OTA	Office of Technology Assessment
PCR	Polymerase Chain Reaction
PS	Producer Surplus
RAC	Recombinant DNA Advisory Committee
rDNA	Recombinant DNA
RED	Registration Eligibility Document
RNA	Ribonucleic Acid
SMS	Safe Minimum Standards
SVWG	Shrimp Virus Work Group
TERA	TSCA Experimental Release Application
TGLO	Texas General Land Office
TSCA	Toxic Substances Control Act
TSS	Total Social Surplus
UK	United Kingdom
UKDOE	United Kingdom Department of the Environment
UKHSE	United Kingdom Health and Safety Executive

USDA	U.S. Department of Agriculture
USEPA	U.S. Environmental Protection Agency
USSD	U.S. State Department
VBNC	Viable But Not Culturable
WRI	World Resources Institute
WTA	Willingness to Accept
WTP	Willingness to Pay

Index

Abalone 51, 68, 74, 92, 157, 163, 180
Abiotic factors 9, 32, 62, 65, 66, 75, 76, 79, 82, 84, 85, 88, 91, 119, 127, 128, 197
Active settlement 50, 74, 87
Advisory Committee on Release to the Environment (ACRE) 14, 26, 27, 29, 221
Aerosols 42, 43
Agricultural Biotechnology Research Advisory Committee (ABRAC) 3, 5, 7, 18, 27, 30, 43, 45, 54, 68, 71, 79, 80, 89, 202, 207, 217, 221
American comb jellyfish 81
American Fisheries Society (AFS) 16, 91, 167, 221
American Petroleum Institute (API) 111, 129, 221
Antifreeze proteins 66, 68, 74, 76, 92, 210
Aquaculture xii, xiii, xiv, 1, 3, 5, 13, 15, 28, 70, 90, 102, 103, 109, 120, 121, 123, 130, 132, 139, 140, 146, 150, 153, 154, 155, 157, 158, 159, 160, 161, 162, 164, 167, 168, 169, 170, 171, 172, 173, 174, 176, 177, 178, 179, 180, 181, 190, 192, 194, 195, 201, 202, 213, 215, 217, 219, 220
Aquatic Nuisance Species 150, 154, 172, 178, 207, 221
    Task Force (ANSTF) 146, 150, 151, 154, 155, 172, 178, 195, 221
Archaea 96, 108, 109, 125
Army Corps of Engineers (COE) 140, 144, 145, 148, 160, 168, 169, 173, 174, 221
Australia xiv, 8, 93, 138

Bacillus thuringiensis (Bt) 20, 97, 109, 110, 132, 134, 135, 136, 221
Bacteria 12, 14, 18, 28, 32, 37, 42, 69, 96, 97, 98, 99, 100, 101, 102, 104, 105, 106, 107, 108, 110, 113, 114, 116, 117, 118, 119, 120, 121, 122, 123, 124, 125, 126, 129, 130, 131, 132, 133, 134, 135, 136, 138, 173, 192, 214, 220
    cyanobacteria 37, 96, 102, 104, 108, 110, 125, 130, 131, 135, 136
    facultative marine 98
    obligate marine 98
Bacteriophage 138
Baltic Sea 37
Barnacles 50, 58, 67
Benefits
    market 186, 187, 198
    net 186, 187, 189
    non-market 197, 200
Benthic layer 195
Berg, Paul vii
Bioaccumulation 111, 113

Bioaugmentation 113
Biodegradation 110, 111, 113, 115, 137
Biodiversity 8, 28, 71, 78, 135, 178, 188, 193, 196, 197, 203, 204, 207, 208
Biofilm 98, 99, 119, 181
Biogeography xiii, 37, 38, 44, 72, 73
Bioremediation xii, xiii, 1, 3, 8, 14, 20, 108, 109, 110, 111, 112, 113, 114, 115, 116, 121, 123, 126, 128, 129, 130, 131, 132, 135, 139, 181, 192, 200, 213, 219, 220
Biotechnology Science Coordinating Committee (BSCC) vii, 2, 3, 29, 221
Biotic factors ix, 32, 38, 43, 52, 65, 76, 79, 82, 85, 88, 101, 119, 127
Bloom 34, 37, 59, 84, 102, 104, 105, 126, 133, 200
Body Size 34, 47, 64, 67, 74, 75, 76, 90, 91, 187, 195

California 11, 89, 91, 157, 162, 163, 165, 168, 171, 179, 180, 207
Canada 8, 41, 85, 120, 121, 122, 125, 128, 131, 137, 153, 190, 196, 198, 207, 209
Center for Public Issues in Biotechnology (CPIB) xii, xv, xvi, 221
Center for Veterinary Medicine (CVM) 5, 7, 29, 159, 161, 169, 179, 221
Chesapeake Bay Program (CBP) 143, 161, 162, 165, 166, 167, 175, 176, 177, 178, 180, 221
Clam 81, 89
Clean Water Act 144, 146, 148, 171, 172, 221
Coastal Zone Management Act 141, 142, 143, 148, 164, 166, 170, 171, 221
Committee on Molecular Marine Biology (CMMB) 101, 134, 221
Commodity 183
Community Nutrition Group (CNI) 16, 28, 68, 90, 157, 178, 221
Competition
  interspecific 55, 77
Congress 3, 111, 114, 137, 149, 154, 158, 159, 160, 178, 180
Connecticut 165, 168, 171
Conservation Law Foundation (CLF) 145, 221

Containment xiv, 7, 9, 10, 15, 18, 21, 22, 23, 24, 26, 27, 31, 68, 70, 71, 72, 73, 86, 117, 118, 126, 127, 132, 136, 139, 157, 182, 201, 202, 203, 204, 206, 215, 216
  biological 117, 118, 136
Convention on Great Lakes Fisheries (CGLF) 153, 221
Coordinated Framework for the Regulation of Biotechnology 2, 218
Copepods 55, 67, 104
Coriolis effect 39
Cost-benefit analysis (CBA) 181, 182, 189, 192, 193, 195, 197, 200, 205, 206, 221
Costs
  abatement 203, 204, 206
  compliance 183, 206, 219, 220
  containment 203, 204
  legal 202
  production 202
  regulatory 160, 202
  travel 200
Currents 36, 37, 39, 40, 41, 42, 43, 47, 50, 58, 59, 72, 76, 82, 117
  transoceanic 38, 48

Damages
  marginal 203, 204
Delaware 112, 137, 143, 146, 168, 171
Demand curve 184, 185, 186, 189, 198
Denmark 14, 26, 27
Department of Commerce (DOC) 140, 142, 152, 158, 174, 180, 221
Department of Interior (DOI) 152, 158, 211, 221
Department of the Environment (UKDOE) 14, 16, 29, 109, 137, 222
Developed countries 193
Developing countries 194
Development
  direct 73
  lecithotrophic 73
  planktotrophic 34, 48, 49, 50, 52, 71, 72, 73, 80, 81, 86
Diatom 83
Dinoflagellates 37, 99, 106, 125
Dioecy 45

Index

Dispersal xii, xiii, xiv, 1, 9, 11, 19, 21, 22, 31, 32, 34, 37, 39, 44, 47, 48, 50, 52, 54, 55, 57, 58, 59, 61, 69, 71, 72, 73, 88, 89, 91, 119, 123, 128, 133, 215, 216, 218
   planktonic 34, 44, 47, 48, 52, 71, 72, 86
   strategies 44, 72
Dissolved oxygen 34, 35, 43, 63
Dolphinfish 70

Ecological analysis model 62, 214
Ecological constraints
   allometric 66, 67, 68, 74, 92
   phylogenetic 66, 67, 74, 132, 134
Ecological effects viii, xiii, 2, 5, 6, 15, 16, 32, 34, 42, 61, 62, 70, 80, 81, 82, 83, 84, 85, 86, 88, 127, 129, 167, 183, 192, 200, 215, 216
   cascading 84, 88
Economics
   ecological xiv, 188, 193, 207
   environmental 188, 192
   neoclassical 193
Ecosystem
   boundaries 32, 43, 44, 52
Ecotourism 194
Eddies 40, 52, 59
Eggs
   demersal 48
   pelagic 35, 41, 42, 48, 49, 54, 55, 56, 58, 60, 84
   size 47, 73, 91
El Niño Southern Oscillation current (ENSO) 38, 221
Endangered Species Act 152, 173
Environment
   coastal xiv, 140, 142
   freshwater xiii, 28, 31, 33, 36, 43, 44, 46, 47, 50, 52, 61, 68, 69, 70, 71, 81, 86, 87, 88, 90, 97, 98, 102, 109, 110, 118, 131, 134, 140, 144, 147, 152, 153, 165, 168, 174, 175, 176, 177, 178
   marine viii, ix, xii, xiii, xiv, 1, 10, 11, 12, 13, 19, 31, 38, 42, 44, 50, 51, 61, 68, 69, 70, 73, 81, 82, 86, 88, 89, 92, 95, 96, 97, 98, 99, 100, 101, 103, 107, 109, 110, 114, 115, 118, 122, 124, 126, 127, 128, 131, 138, 139, 140, 141, 144, 145, 146, 147, 148, 152, 153, 157, 161, 164, 165, 168, 169, 170, 176, 192, 195, 202, 214, 215, 217
   terrestrial xii, 1, 4, 6, 7, 12, 13, 32, 69, 88, 95, 99, 103, 109, 118, 120, 122
Environment Canada 128, 131
Environmental Protection Agency (EPA) vii, viii, xiii, xvii, 2, 3, 4, 5, 6, 7, 14, 15, 19, 20, 23, 24, 25, 30, 114, 115, 120, 121, 122, 123, 125, 127, 128, 137, 143, 144, 145, 146, 147, 148, 217, 218, 221, 223
Escape viii, xii, 16, 23, 62, 63, 64, 65, 66, 67, 68, 69, 70, 71, 72, 73, 74, 77, 78, 79, 81, 82, 85, 86, 87, 88, 116, 127, 161, 163, 172, 187, 194, 201, 202, 203, 204, 205, 215, 216
   probability of xii, 62, 64, 65, 66, 67, 69, 73, 86, 88, 216
Establishment viii, 29, 62, 67, 73, 74, 75, 76, 77, 78, 79, 80, 81, 86, 87, 143, 172, 201, 202, 206
   probability of 73, 74, 75, 76, 77, 78, 79, 80, 86, 87, 88
   rate of 62, 81, 86, 87, 88, 216
European Union xi
Exclusive Economic Zone (EEZ) 141, 153, 174, 218, 221
Executive Order 11987 151, 155, 172
Exotic species 11, 64, 71, 82, 83, 90, 92, 143, 149, 150, 151, 152, 153, 154, 156, 165, 172, 174, 177, 178, 193, 195, 215
Exposure 5, 6, 7, 18, 24, 25, 26, 37, 86
Externality 183, 188, 189, 190, 193, 197, 200
Extremophile 96

Familiarity 6, 8, 115, 116, 118, 126, 127, 128, 129
Fecundity 18, 45, 46, 47, 52, 60, 67, 70, 73, 74, 81, 87, 91, 122
Federal Food, Drug and Cosmetic Act (FFDCA) 3, 159, 161, 217, 221
Federal Insecticide, Fungicide, and Rodenticide Act (FIFRA) 6, 20, 221

Field test 8, 24, 26, 28, 29, 63, 70, 79, 90, 109, 115, 116, 126, 127, 161, 163
Final Rule for Microbial Biotechnology Products 4
Fish and Wildlife Coordination Act 144, 173, 174, 221
Fish and Wildlife Service (FWS) 144, 147, 150, 151, 152, 155, 173, 174, 178, 180, 194, 200, 207, 211, 217, 221
Fisheries 16, 83, 91, 102, 107, 142, 143, 148, 158, 161, 164, 167, 170, 173, 174, 188, 190, 192, 194, 195, 201, 206, 211, 213, 217
Fishing industry 190
Florida 85, 90, 130, 132, 141, 165, 168, 171, 209
Flounder 46, 66, 68
Food and Agricultural Organization (FAO) 190, 192, 196, 198, 208, 221
Food and Drug Administration (FDA) viii, xiii, xvii, 2, 3, 5, 7, 15, 23, 24, 29, 156, 157, 159, 161, 169, 217, 221
Food webs 32, 52, 131, 215
Foodfish 194, 204
Fungi 96, 103, 109, 131

Gaps in the science base xi, 1, 9, 17
Genes 13, 20, 66, 67, 76, 101, 107, 109, 110, 113, 117, 119, 120, 124, 125, 127, 128, 135, 149, 167, 173
Genetic engineering vii, viii, 9, 67, 91, 96, 107, 108, 110, 113, 173, 180, 209, 215
Genetically engineered organism (GEO) xi, xii, xiii, xiv, 1, 3, 4, 5, 6, 9, 10, 11, 12, 13, 14, 15, 16, 17, 19, 31, 34, 62, 63, 66, 67, 72, 73, 74, 75, 77, 78, 80, 81, 82, 83, 85, 87, 88, 139, 140, 147, 148, 149, 150, 152, 153, 154, 155, 156, 159, 163, 165, 166, 167, 168, 170, 174, 176, 181, 182, 183, 187, 188, 189, 190, 192, 193, 194, 195, 196, 197, 198, 200, 201, 202, 203, 204, 205, 206, 207, 214, 215, 216, 217, 218, 222
Georgia 165, 168, 171
Geothermal vents 36
Great Lakes 12, 33, 71, 90, 91, 123, 147, 153, 164, 195, 221, 222

Growth rate 64, 67, 68, 70, 74, 182
Gyres 40, 41, 52

Habitat 10, 12, 42, 43, 44, 51, 60, 62, 63, 65, 66, 69, 70, 72, 73, 74, 75, 76, 77, 78, 79, 80, 81, 83, 84, 85, 87, 99, 120, 143, 152, 174, 182, 216
  exotic 63, 87
Haploid 79
Hazard 5, 6, 7, 17, 25, 86, 105
Hermaphroditism
  sequential 45, 78
  simultaneous 45, 78
Human health xii, 2, 6, 18, 110, 114, 118, 126, 127, 132, 150, 215, 217, 219
Hybrid 73, 78, 93, 177
Hydrostatic pressure 33, 34

International Committee for the Exploration of the Seas (ICES) 11, 14, 15, 26, 27, 222
Introduction
  accidental 31, 62, 63, 70, 71, 129, 195
  planned 61, 72, 73, 93, 210
Invasion ecology 81
Isopods 67

Lacey Act 149, 150, 155, 164, 169, 173, 180
Larva size 46, 71
Life history strategies xiii, 31, 32, 33, 35, 44, 47, 61, 63, 68, 69, 74, 83, 87
Louisiana 165, 168, 171

Magnuson Fishery Conservation and Management Act 140, 142, 164, 174
Maine 41, 164, 165, 168, 171
Marine ecosystem services 192, 196
Marine Protection, Research, and Sanctuaries Act 147
Market effects 183
Market goods 188, 192
Maryland 106, 143, 162, 165, 167, 168, 171, 172, 178
Massachusetts 54, 55, 142, 157, 162, 165, 168, 171, 179, 187
Meiofauna 96, 104, 109
Mesocosm 101, 118, 119, 120, 133

Index 229

Mexico 59, 103, 112, 123, 133, 142, 153, 209
Microcosm 21, 118, 119, 120, 134, 137
Mississippi 162, 164, 165, 166, 171, 209
Molecular phylogenic analysis 97
Mortality 34, 36, 44, 67, 68, 73, 74, 76, 85, 87, 93, 107, 135, 211
Mussels 11, 67, 93, 153, 166, 195, 211

National Academy of Sciences (NAS) 6, 8, 29, 60, 61, 91, 136, 222
National Aquaculture Act 158
National Environmental Policy Act 145, 146, 148, 155, 165, 170, 222
National Institutes of Health (NIH) vii, viii, 2, 3, 21, 162, 163, 222
National Invasive Species Act 153
National Marine Fisheries Service (NMFS) xvii, 133, 142, 144, 146, 148, 155, 169, 174, 178, 180, 192, 209, 217, 222
National Oceanographic and Atmospheric Administration (NOAA) xii, 142, 147, 159, 170, 174, 178, 209, 217, 222
National Science Foundation (NSF) iv, 59
Native species 6, 64, 74, 75, 82, 83, 84, 133, 139, 151, 165
Natural selection 44, 67, 68, 75, 150
New York 162, 165, 171, 179
Niche
  ecological 11, 63, 64, 65, 66, 67, 75, 76, 91, 92
  fundamental 64, 65, 66, 75, 76, 216
  realized 64, 65, 66, 75
Nonindigenous Aquatic Nuisance Prevention and Control Act 153
North American Wetlands Conservation Act 147
Nudibranches 81

Occupational Safety and Health Administration (OSHA) iv
Office of Science and Technology Policy (OSTP) vii, 2, 17, 29, 210, 222
Organisms
  exotic 10, 118, 151, 156
  macroorganisms viii, xiii, 10, 14, 35, 36, 61, 95, 98, 99, 102, 104, 105, 107, 120, 121, 122, 195

microorganisms viii, xii, xiii, 8, 13, 20, 23, 25, 28, 29, 32, 35, 36, 37, 41, 43, 84, 95, 96, 97, 98, 99, 102, 104, 107, 108, 111, 112, 114, 115, 116, 117, 118, 119, 120, 121, 122, 123, 125, 126, 127, 128, 129, 130, 131, 132, 135, 136, 137, 139, 173, 193, 213, 214, 215, 217
transgenic viii, xii, xiii, 2, 4, 8, 9, 11, 13, 14, 15, 16, 17, 19, 31, 32, 52, 61, 62, 68, 69, 71, 72, 73, 74, 78, 79, 80, 81, 82, 83, 84, 87, 88, 92, 95, 140, 141, 142, 143, 146, 147, 148, 149, 150, 152, 154, 157, 158, 159, 160, 164, 165, 169, 170, 176, 180, 181, 182, 201, 202, 215
Organization for Economic Cooperation and Development (OECD) 8, 28, 29, 222
Overfishing 190, 195, 196, 206

Parasites 10, 21, 27, 76, 106
Parent species 5, 64, 65, 66, 68, 73, 74, 75, 78, 166, 182, 190
Pathogenicity 21, 23, 43, 63, 107, 115, 120, 121, 122, 124, 125, 130, 134
Pfisteria 106
Pharmaceuticals xi, xii, 220
Phenotype 6, 10, 12, 13, 63, 64, 66, 67, 100, 150
Photic effects 33, 34, 35, 42
Plankton 30, 37, 41, 48, 49, 55, 72, 96, 97, 104, 105, 106, 107
  bacterioplankton 49, 98, 105
  holoplankton 49
  meroplankton 48
  phytoplankton 34, 37, 42, 43, 45, 49, 51, 55, 56, 57, 58, 59, 83, 84, 105, 136, 137, 138, 193, 200
  zooplankton 42, 49, 51, 63, 84, 85, 93, 102, 105, 106, 123, 125, 136
Plasmid 101, 119, 120, 131, 132, 133, 135, 136
Points to Consider (EPA) 4, 5, 7, 10, 12, 14, 15, 18, 20, 30, 115, 218
Polymerase chain reaction (PCR) 101, 135, 222
Population 8, 35, 38, 54, 55, 63, 64, 71, 73, 74, 76, 78, 80, 81, 82, 84, 85, 86,

93, 117, 121, 183, 190, 194, 204, 211, 213
  transgenic xiii, 62, 81
Predator 13, 31, 48, 64, 65, 75, 76, 83, 84, 85, 92, 130, 152, 204
Presidential/Congressional Commission on Risk Assessment 15, 18, 29
Prey 13, 21, 31, 64, 66, 67, 76, 77, 82, 85, 149, 193, 216
Pricing
  hedonic 200
  implicit 200
Primary production 40, 44
Propagules 38, 48, 49, 50, 52
Protozoa 96, 105, 106, 109, 110, 135
Pseudomonas 95, 98, 112, 114, 116, 117, 118, 119, 124, 125, 126, 134
Public policy 28, 29, 182

Rate of spread 81, 82, 86, 87, 88, 216
Recombinant DNA (rDNA) vii, xi, 3, 117, 133, 136, 162, 163, 222
Recombinant DNA Advisory Committee (RAC) 3, 222
Red tide 85, 130
Regional Fishery Councils 142
Regulatory reform 214
Reproduction 1, 9, 21, 44, 45, 51, 52, 61, 67, 68, 76, 78, 80, 106, 215, 216
  asexual 45, 78
  modes of 45, 74
  success 16, 67, 68, 78
Resources
  novel 63, 64, 65, 66, 67, 73, 75, 76, 83, 182, 216
Rhode Island 165, 171
Rings
  cold-core 40, 58
  warm-core 40
Risk
  assessment viii, ix, xi, xii, xiii, 1, 2, 3, 4, 5, 6, 7, 8, 9, 10, 11, 12, 13, 14, 15, 16, 17, 18, 19, 29, 31, 61, 68, 69, 70, 72, 73, 75, 81, 85, 86, 88, 89, 92, 96, 98, 124, 125, 127, 129, 134, 156, 158, 164, 165, 167, 169, 170, 180, 181, 183, 193, 202, 206, 214, 217, 218

  ecological viii, xii, xiii, 2, 5, 61, 62, 64, 66, 70, 73, 79, 80, 86, 87, 88, 95, 96, 100, 109, 159, 161, 169, 181, 182, 214, 215, 216, 217, 219
  management xiii, 2, 3, 4, 16, 62, 81, 182, 217
  risk assessment protocols xii, xiii, 1, 2, 5, 6, 7, 8, 9, 10, 11, 13, 14, 15, 16, 18, 19, 20, 24, 26, 27, 79, 85, 86, 154, 155, 158, 164, 165, 168, 175, 177, 182, 202, 206, 214, 217, 218, 219
Rivers and Harbors Act 144

Safe minimum standard (SMS) 196, 197, 206, 222
Salinity 18, 36, 43, 63, 66, 76, 98, 101, 136, 137
Salmon 5, 18, 29, 36, 38, 47, 66, 68, 71, 74, 76, 77, 78, 84, 90, 91, 93, 149, 157, 160, 169, 180, 187, 190, 196, 197, 198, 199, 204, 208
  aquaculture 190
  Atlantic 66, 68, 74, 75, 76, 90, 92, 198, 210
  chinook 68, 146
  coho 29, 68, 196
  Pacific 41, 58, 68, 74, 187, 196
  pink 71, 90
Sargasso Sea 40, 56
Seafood demand 190, 192
Seastars 84, 91
Shellfish viii, xii, 3, 13, 28, 63, 64, 66, 79, 89, 95, 103, 104, 105, 107, 125, 137, 156, 157, 159, 161, 168, 174, 175, 182, 213, 215, 217, 219
South Carolina 165, 168, 171
Sport fishing 194
Sterility 79, 89, 201, 202
  induced 74, 79, 80, 202, 216
Stratification 37
Substrate 34, 76, 82, 100, 137, 169
Suicide construct 117
Supply curve 185, 186, 187, 189
Surplus
  consumer 183, 184, 197, 198, 199, 200
  producer 185, 186, 198
  social 186, 198

Index

Sybron Biochemical 111, 112, 136

Temperature gradients 32, 38, 42
Texas 112, 137, 141, 164, 165, 168, 171, 222
Tide 37, 41, 43, 47, 52, 56, 106, 144, 200
Toxic Substances Control Act (TSCA) 6, 14, 19, 20, 30, 128, 135, 217, 222
Toxin 8, 20, 85, 98, 104, 106, 109, 110, 117, 120, 121, 125, 132
Transduction 99, 100
Transformation 37, 99, 100, 132, 133, 193
Transgene 73, 78, 85, 87, 202
Transgenic induction 63, 64, 65, 66, 68, 70, 73, 74, 75, 76, 77, 87
Triploid 79, 80, 89, 91, 166, 175, 202
Trophic level ix, 51, 82

U.S. Department of Agriculture (USDA) vii, viii, xiii, 2, 3, 4, 5, 7, 14, 22, 23, 25, 26, 27, 29, 30, 70, 93, 152, 155, 158, 159, 202, 207, 211, 217, 223
Uncertainty xii, 95, 127, 145, 150, 197, 214, 215
United Kingdom 14, 15, 26, 27, 28, 29, 30, 55, 131, 132, 135, 137, 190, 221, 222
United Nations World Food Conference vii
Unpriced by-products 188
University of Maryland Biotechnology Institute xii, xv, xvi, xvii
Upwelling 38, 42, 51, 54, 56, 59, 60

Valuation 188, 192, 197, 200, 201, 206
Viable but nonculturable (VBNC) 101, 126, 223
Vibrio
  Vibrio cholerae 102, 130, 132, 136, 138
  Vibrio vulnificus 132, 134
Virginia 137, 143, 160, 162, 165, 168, 171, 179
Virus viii, x, 12, 19, 22, 30, 34, 96, 100, 101, 106, 107, 109, 120, 123, 132, 135, 136, 138, 173

Washington 38, 83, 84, 143, 165, 166, 168, 171
Washington, D.C. 143
Water
  brackish 147
  chemistry 33, 58
  density 5, 33, 40, 50
  movement 31, 32, 72, 105
  salt 33, 36
  temperature 32, 37, 38, 41
  viscosity 33, 34
Waves 43, 48, 54, 55, 56, 58, 60, 104
Wildlife x, 140, 143, 144, 147, 149, 150, 151, 155, 161, 162, 164, 166, 173, 177
Willingness-to-accept (WTA) 201, 223
Willingness-to-pay (WTP) 201, 223
Wind 21, 36, 37, 39, 42

Zebra mussels 195